Encyclopaedia of

Sustainable Water Resource Management

Encyclopaedia of

Sustainable Water Resource Management

Volume 3

Editor:

R.N. Sahay

AGROTECH PRESS

JAIPUR ▪ NEW DELHI

AGROTECH PRESS

Regd. Off. "Bony Residency", Gate No. 2
Opp. Tilak Public School, Vishweswariya Nagar,
Gopalpura Road, Jaipur - 302018, India
Ph: 0141 - 2761280, Telefax: 2761381
E-mail : agrotechpress@gmail.com

Delhi Office: 102, 1st Floor, Satyam House,
4327/3, Ansari Road, Darya Ganj,
New Delhi - 110002, India
Ph: 011-45652440
visit us: www.agrotechpress.com

Distributed By
KEYA INTERNATIONAL PRESS LLP
E-mail: keyainternationalpressllp@gmail.com

Published in India, 2025

ISBN : 9789384568269 (5 Vols. Set)

Typeset by Thomson Press

Printed and bound in India by Thomson Press Ltd.,
New Delhi

Contents

Preface *vii*

1. **Maintaining the Status Quo: Protecting Established Water** 1

Water Use in the Northwest • The Traditional Basic Rules of Western Water Law • The Status Quo Policy in the Northwest • Enforce the Law only When Necessary • Effects of the Status Quo Policy • Implications of the Status Quo Policy

2. **New Generation of Indicators for Coastal Waters** 22

Limitations of Current Coastal Indicators • Multiple Stressors of Environmental Condition • Biological Community Responses as Condition and Change Indicators

3. **Water Rights: The Water Policy Implications** 36

Water Development and Management on the Columbia • The National Academy of Sciences Study • Water Policy Implications of the NAS Study • The Electricity-water Nexus and the Electric Utility Sector • The Electricity Water Nexus • Electricity Water Challenges in Eight Metropolitan Areas • Electric Utilities and Regulators

4. **Anthropology and Policy in Global Context** 92

The Research-policy Cycle • Anthropology and the Research-policy Cycle • Anthropologists as Participants in the Policy Arena • Water as Local Resource, Water as Global Resource • Water Markets • Water Banks • Sustainable Management • Reorienting the Debate: Three Crises Rolled into One • Clarifying the Global Dimensions of Water Scarcity • Global Awareness and Global Solutions

5. **Irrigation Water Deliveries versus Consumptive Use** 125

Previous Analyses of Price-Responsiveness of Deliveries • Modelling Procedure • Policy Considerations • Effects of Threshold Drainage Area • Flood Hazard Modelling

6. **The Drainage Services** 149

The Supply of Drainage Services • Drainage Water Quality • Case

Study: Farmland Drainage in Norfolk, UK • Drainage Needs in the Lower Indus, Pakistan • Public Goods and the Christian Eucharist • Institutional Economics and Irrigation Policy

7. Watershed-based Integration of Hydrology **174**

Field Sites and Instrumentation • Use of the Watershed in Hydrology Courses • Use of the Watershed in Hydrogeology • Use of the Watershed in Environmental Geophysics • Use of the Watershed in Environmental Measurements • Transferability • Watershed-based Summary Field Exercise • Study Area Setting • Activities • Instructional Effectiveness and Limitations

8. Water Quality Testing of High-Use **191**

Materials and Methods • Organisation Theory: Analysis of Water Management • Modern Organisation Theory • Policy Narrative • Eurostories • Water as a God-given Right • Water as a Commodity • The Response by Organisation Theory • The Prospects for the Beneficial Use Doctrine to Achieve Efficiency in Western Water Use • The Management of International Waters: A Global Policy • The Problem • Global Policy Discourse on Integrated Water Resources Management

9. Rare Resources and Crises **222**

The Context of Research: South-Eastern Morocco and the Dra Valley • The Place of Ait Unzar in South-Eastern Morocco • A Case Study: The Ait Unzar Nomads and the Village of Tiraf • The Persistence of an Overarching 'Tribal Identity' • Antimicrobial Resistance in Escherichia Coli Isolated in Wastewater • Wastewater Treatment Plants • Sample Collection • Results • Cryptosporidium Inactivation by Low-Pressure UV in a Water Disinfection Device • Cryptosporidium and UV Disinfection • Oocyst Exposure to UV and Process Control • Infectivity Calculations

10. Outbreaks in Drinking-Water Systems **242**

Outbreak Statistics • Water System Deficiencies • Distribution System Deficiencies • Outbreaks of Undetermined Etiology • Water System • Coliforms in Water During Outbreaks • Diarrheal Diseases in Children from a Water Reclamation • Data Management and Analysis

11. The Challenges of Sustainable Access to Safe Drinking Water **258**

Water Quality • Water Coverage, Quantity, Continuity, and Cost • Holy and Wholesome • The Global Water Crisis • Faith and Water • Liturgical Theology • Liturgical Praxis and the Water Crisis • Holy Water—Please Do Not Drink! • Come to the Waters—Access to Water and Inequality

Bibliography *277*

Index *279*

Preface

Water resources are sources of water that are useful or potentially useful. Uses of water include agricultural, industrial, household, recreational and environmental activities. Virtually all of these human uses require fresh water. 97% of the water on the Earth is salt water. However, only three percent is fresh water; slightly over two thirds of this is frozen in glaciers and polar ice caps. The remaining unfrozen freshwater is found mainly as groundwater, with only a small fraction present above ground or in the air.

Fresh water is a renewable resource, yet the world's supply of clean, fresh water is steadily decreasing. Water demand already exceeds supply in many parts of the world and as the world population continues to rise, so too does the water demand. Awareness of the global importance of preserving water for ecosystem services has only recently emerged as, during the 20th century, more than half the world's wetlands have been lost along with their valuable environmental services for Water Education. The framework for allocating water resources to water users is known as water rights. Water management is the activity of planning, developing, distributing and managing the optimum use of water resources. In an ideal world, water management planning has regard to all the competing demands for water and seeks to allocate water on an equitable basis to satisfy all uses and demands. This is rarely possible in practice. Half of the world's people now live in towns and cities, a figure expected to reach two-thirds by 2050. In the areas surrounding urban centres, agriculture must compete with industry and municipal users for safe water supplies, while traditional water sources are becoming polluted with urban wastewater. As cities offer the best opportunities for selling produce, farmers often have no alternative to using polluted water to irrigate their crops. Depending on how developed a city's wastewater treatment is, there can be significant health hazards related to the use of this water. Wastewater from cities can contain a mixture of pollutants. There is usually wastewater from kitchens and toilets along with rainwater runoff. This means that the water usually contains excessive levels of nutrients and salts, as well as a wide range of pathogens. Heavy metals may also be present, along with traces of antibiotics and endocrine disruptors, such as oestrogens. Developing world countries tend to have the lowest levels of wastewater treatment. Often, the water that farmers use for irrigating crops is contaminated with pathogens from sewage. The pathogens of most concern are bacteria, viruses and parasitic worms, which directly affect farmers' health and indirectly affect consumers if they eat the contaminated crops. Common illnesses include diarrhoea, which kills 1.1 million people annually and is the second most common cause of infant deaths. Many cholera outbreaks are also related to the reuse of poorly treated wastewater. Actions that reduce or remove contamination, therefore, have the potential to save a large number of lives and improve livelihoods. Scientists have been working to find ways to reduce contamination of food using a method called the 'multiple-barrier approach'. One of the biggest concerns for our water-based resources in the future is the sustainability of the current and even future water resource allocation. As water becomes more scarce the importance of how it is managed grows vastly. Finding a balance between what is needed by humans and what is needed in the environment is an important step in the sustainability of water resources. Attempts to create sustainable freshwater systems have been seen on a national level in countries such as Australia and South Africa, and such commitment to the environment could set a model for the rest of the world.

The field of water resources management will have to continue to adapt to the current and future issues facing the allocation of water. With the growing uncertainties of global climate change and the long term impacts of management actions, the decision-making will be even more difficult.

This book tries to asses water resources and then utilisation. It discuses the water resource potentiality and its utilization pattern, problem of water quality, water supply for domestic use and agriculture, impact of droughts on water supply, planning and management for sustainable development of water resources in rural areas. This book is useful for public health engineering departments, state and central ground water board, agricultural departments, planners, non-government organisations, research scholars of geography, geology and other allied disciplines.

—Editor

1

Maintaining the Status Quo: Protecting Established Water

For decades, water law in all four Pacific Northwest states has been based on the prior appropriation doctrine. That doctrine has been a fixture in the western United States for over a century, and the fundamental rules of water law based on prior appropriation are well established. These basic rules provide water users with a high degree of certainty and security, creating private property rights to use a resource that is owned by the public, but they also limit water use in some significant ways.

For example, the traditional rules restrict where, when, how, and how much water may be used, and specify how water rights may be established and lost. The Northwest's water resources and those who rely on them have come under increasing stress in recent years. Causes of this stress include overappropriation, drought, population growth, the decline of salmon and other fish populations, aquifer depletion, water quality impairment, assertion of tribal water rights, increasing competition for water supplies, public demands for environmental protection, and other factors. These factors have helped create and exacerbate water conflicts, often forcing the states to face difficult issues they had long avoided.

In facing these issues, the states commonly have not applied the traditional rules of western water law. Instead, states often have effectively waived or abandoned these rules in order to preserve existing water-use practices. State deference to existing water uses takes many forms, from silent yet unmistakable failures to enforce long-standing rules, to changes in state statutes for the express purpose of allowing established uses to continue. The accommodation of status quo water uses is now the overarching principle of Northwest water policy. Simply put, the Northwest states protect water users' established practices more faithfully than their legal rights. Where existing water uses are inconsistent with traditional rules of water law, the states have often bent, changed, or ignored those rules in order to preserve these established practices.

This unwritten policy favours those appropriators whose uses would be curtailed by a straightforward application of water law principles and disadvantages others, including other existing water users, prospective new users, and those interested in restoring instream flows, who would benefit if these principles were implemented. Many of the traditional rules of western water law are fading in practical importance as legislatures, courts, and agencies protect existing water uses from the application of these rules. The states' policy of protecting existing uses seems to be based on considerations of politics, economics, and equity. But this policy may damage other water users, impair instream flows, limit economic opportunities, and ultimately create greater uncertainty about water use and management.

Water Use in the Northwest

People not familiar with the Pacific Northwest, and even some who are, often believe the region is far wetter than it really is. The region's legendary rainfall is not evenly distributed, and the area east of the Cascades is typically quite dry. In Idaho, average annual precipitation in the twenty inches. Even west of the Cascades, where winter precipitation is generally abundant, many streams run short of water in the summer and fall. In the Dungeness and Quilcene basins on Washington's Olympic Peninsula, for example, low streamflows are a chronic problem. Given the demands for water to meet existing and new uses, both instream and out-of-stream, most of the Northwest does not have enough water.

Irrigation is, by far, the dominant out-of-stream water use in the region. Irrigation accounts for over eighty-seven percent of total water withdrawals in the Northwest and nearly ninety-seven percent of the region's water consumption. The next largest water use, public supply, accounts for less than five percent of the withdrawals, while industrial withdrawals are less than three percent. The percentages vary somewhat by state. In Idaho, for example, irrigation accounts for ninety-five percent of water withdrawals and livestock use exceeds public supply and industrial withdrawals combined. In Washington, the most urbanised of the Northwest states, public supply makes up eleven percent of water withdrawals and industrial use more than six percent, but irrigation still exceeds seventy-six percent of total withdrawals. Water withdrawals have sharply depleted stream flows in many Northwest rivers. As in other parts of the West, state efforts to protect instream flows have been too little and too late to keep rivers from drying up. Also, water diversions and low streamflows have seriously harmed fish and other aquatic and riparian organisms. Inadequate instream flows related to water withdrawals are a continuing and significant problem for western rivers and aquatic ecosystems. These water use patterns of the Northwest are not a recent development. They have been firmly established for decades, within the framework of state water laws based on the bedrock principles of the venerable prior appropriation doctrine. The following subsection briefly examines some of these principles.

The Traditional Basic Rules of Western Water Law

Western water law under the prior appropriation doctrine essentially is based on a handful of fundamental and long-standing rules. These few traditional rules are clearly established and simple to state, if not to apply. In general terms, these rules follow.

State Control of Waters

Perhaps the most basic rule of western water law is that the states control the use of water, which state laws declare to be a public resource. States provide for private rights to use water, but those rights may be established only as authorised by the states. Moreover, exercise of such rights is subject to state administration and enforcement.

State Approval of Water Rights

The traditional steps needed to establish a water right are intent, diversion of water from its natural source, and application of water to a beneficial use. Today, however, new appropriations may be made only on the basis of a permit issued by a state water resource agency.

First in Time, First in Right

A water right's priority is based on the date of appropriation, and the oldest rights have the highest priority. Where water is insufficient to satisfy all rights, junior appropriators' withdrawals are curtailed so that users with senior rights may continue to receive their full water supply.

Beneficial Use Without Waste

Water may be appropriated only for a specified "beneficial use," and water rights authorise the use only of enough water to satisfy that beneficial use. In addition, water uses must be reasonably efficient; no one has a legal right to "waste" water.

Appurtenancy

Water rights usually specify a particular place, as well as type, of beneficial use. In most states, water rights are appurtenant to a specific parcel of land, and may not be used elsewhere without a transfer.

Conditional Transferability

The holder of a water right may change its point of diversion, place, or purpose of use, but only with prior state approval. Any such change, however, must not injure the rights of other appropriators, whether junior or senior to the right being transferred.

Use it or Lose it

A water user may lose her water right by failing to exercise it, either by nonuse of water for a period of years (forfeiture) or nonuse coupled with evidence of intent to abandon the right (abandonment). The foregoing list does not attempt to identify all the rules of western water law, or provide a detailed analysis of individual rules. It does, however, provide a brief summary of the most fundamental and familiar points of the prior appropriation doctrine. In theory, these traditional rules form the basis of water law in the Northwest states. As the following section explains, however, these rules are often sacrificed for the protection of existing water uses.

The Status Quo Policy in the Northwest

The Northwest states remain officially committed to the prior appropriation doctrine. In reality, however, the states often effectively waive the traditional rules where they threaten to curtail established water uses. States have routinely bent, changed, or ignored the traditional rules, through action or inaction, in order to maintain such uses. As a practical matter, water management in the region today is based less on strict prior appropriation principles than on protection of the status quo, that is, preservation of the water use practices, legal or not, that have become established in an area.

"Where is the inconsistency?" the reader may ask. "Isn't prior appropriation itself based on protection of existing uses?" Certainly the traditional rules of western water law protect these uses in many ways, but the rules also establish limits to that protection.

For example, users who violate the limits of their water rights by exceeding the specified rate, duty, or season of use, or applying water to unauthorised lands, are not protected by the traditional rules. Nor are users protected who fail to use their water rights for an extended period, who waste water, or who do not observe legal requirements for establishing or maintaining a water right. Junior appropriators are not protected from being shut off in favour of seniors during times of shortage, even if the juniors have not previously been subject to such regulation. And the traditional rules do not prevent water from being used in new places or for new purposes, so long as such changes injure no other appropriators and are approved by the state. Clearly then, the traditional rules of western water law do not provide absolute protection to existing water uses.

Two points must be noted at the outset of this discussion of the states' practice of protecting established water uses. First, there is no inherent conflict between preserving the status quo and applying traditional water laws. To the contrary, maintaining the status quo is fully consistent with prior appropriation rules so long as 1) all users' rights are clearly established, 2) water has been consistently well managed (based on good data) and distributed among those users by priority, and 3) all appropriators are using water in accordance with the terms of their rights and with reasonable efficiency. However, to the extent that these conditions are not fully met, then current practices may deviate from the law. Where water rights are somehow in question, good data are lacking on basin hydrology or actual water use, or there is inadequate water use regulation, then existing uses may differ dramatically from users' legal rights. Depending upon the circumstances, some users will be better off with the status quo, while others will benefit if water is used and managed "by the book."

Second, while the Northwest states usually preserve the status quo at the expense of water law principles, there are exceptions. The results are not entirely consistent largely because state governments are not monolithic. State water-resource agencies, citizen commissions and boards, legislatures, governors, trial courts, and appellate courts all have jurisdiction over water matters. Often, one state government entity has sought to preserve an established water use by changing or not applying the law, while another entity of the same state government has tried to uphold the law. For example, the governors of both Oregon and Washington have vetoed recent legislative efforts to revise state law in favour of status quo water uses. When the Idaho Department of Water Resources refused to curtail groundwater use to protect senior appropriators, the state Supreme Court held that the agency had a mandatory duty to regulate. In Washington, when the Department of Ecology took this same action, the state supreme court held that the agency had no authority to do so. Given these common divisions within a state government, it is somewhat imprecise to speak of "state" efforts to protect established water uses, and defenders of the status quo do not always have the last word.

Having stated these caveats, this Chapter examines the recent actions and practices of Northwest states with respect to established water uses. The record shows that the states commonly pursue an informal, unofficial policy of protecting the water use status quo even when it conflicts with traditional water law principles. This status quo policy, like prior appropriation, has a few fundamental rules. Unlike the basic features of prior appropriation, however, these rules are difficult to locate in any treatise or casebook on western water law. Instead, they must be synthesized from state efforts to protect existing uses. With rare exceptions, states abide by the following basic precepts: 1) enforce the law only when necessary, 2) change the law where needed to protect existing uses, 3) avoid the position of having to curtail established water uses, and 4) prevent instream demands from threatening existing out-of-stream uses. This section discusses these basic tenets of the status quo policy and examines how states have implemented them.

Enforce the Law only When Necessary

In deciding whether to curtail ongoing water uses, the water resource agencies in the Northwest states take a passive approach. If "enforcement" is defined as requiring water users to comply with applicable laws and water right conditions, such as duty or place of use, all four states take enforcement action against water users almost exclusively in response to complaints from other users. Montana's official policy is that a "complaint is always required unless the regional manager can document compelling reasons for recommending enforcement without one." Even when disputes arise between groups of existing users, the agencies can be reluctant to step in.

In Whatcom County, Washington, the state Department of Ecology (Ecology) has taken little action against widespread illegal water use, despite a 1993 survey that found over five hundred users taking water without a valid right. Most of these users were irrigators, and many had been using water illegally for decades, without a valid state permit, decreed right, or precode claim. Despite this clear violation of state law, Ecology took no meaningful enforcement action against the Whatcom County users. Washington Governor Gary Locke vetoed a bill approved by the 1997 Washington legislature that would have granted amnesty to the Whatcom County users. Locke did, however, sign a bill that allowed these users to file water right claims that they had already relinquished under state law, effectively authorising the uses to continue.

It is impossible to ascertain the true extent of noncompliance with water laws and water rights in the Northwest, largely because the state water resource agencies do not actively seek out violators. A perception exists that violations are common, and that enforcement is infrequent or ineffective. For example, the Snake River Salmon Recovery Team stated "[t]he dewatering of salmon breeding and rearing habitat must be eliminated. The monitoring of screen use, screen performance, and the quantity of water diverted must be greatly improved." But most of the information is anecdotal, and there has been no systematic study of water law violations and enforcement.

Recent events in the Wallowa River Basin of northeast Oregon illustrate the problems of ascertaining noncompliance. An anonymous writer claiming to work for the Oregon Department of Fish and Wildlife sent a series of letters to state officials and a Portland television station alleging widespread and serious water right violations in the Wallowa Basin. The letters claimed that data obtained from state sources showed actual diversions often far exceeding legal limits. The Oregon Water Resources Department (OWRD) responded that these allegations were based on mistaken assumptions about the limits of Wallowa Basin water rights, and denied that there was significant illegal use in the area. Department officials acknowledged, however, that they had very limited enforcement staff in the Wallowa Basin since the basin's watermaster position had been discontinued in the 1980s. Given this lack of staff, the Department itself may not know whether all water users are in compliance.

The environmental group WaterWatch approached state officials in 1996 with concerns about excessive water diversions in the Wood River Basin of southern Oregon, but the Water Resources Department essentially denied that illegal diversions were occurring. WaterWatch then employed a hydrologist who used a flow metre to take measurements of Wood River diversions. The hydrologist reported back to the state that there did appear to be serious illegal water use in that basin. State officials continued to dispute the WaterWatch findings, but acknowledged that "the concerns raised by WaterWatch have some merit. We believe, as does WaterWatch, that the lack of adjustable headgates in some cases, or lack of easily read measuring devices in other cases, can result in excessive diversions" Shortly thereafter, the agency ordered Wood River irrigators to install adequate headgates.

Oregon law actually authorises the Water Resources Department to allow illegal water uses to continue temporarily. Specifically, the Water Resources Director may issue a "limited license," a sort of revocable temporary water right, in connection with an enforcement: order to address an illegal water use. Essentially, the law authorises the director to condone an otherwise illegal water use, provided the director orders the user to comply with the law within a specified period of time.

The Montana Department of Natural Resources and Conservation (DNRC) has an unofficial policy of deferring to informal arrangements (such as water-right pooling or rotation agreements) among local water users on a stream, even when those arrangements are "extralegal." DNRC staff believe that such arrangements are common in Montana. DNRC not only tolerates these "extralegal" arrangements, but actually supports them, primarily because DNRC believes they are usually well tailored to the hydrologic conditions of the watershed, and local water users like them.

None of the Northwest states have taken strong action to implement the requirement of beneficial use without waste. As one commentator recently concluded: The law in Idaho, Montana, Oregon and Washington requires that water not be wasted but there has been no meaningful enforcement of this requirement in any of the states. In all four states, the failure to enforce appears to be because the states lack 1) information on actual water use, 2) a clear waste definition, and 3) political support or wherewithal for anti-waste enforcement. The consequences of this failure include injury to other legal water right holders, both instream and out-of-stream, as well as harm to the public's rights as the owners of the water resource.

When the Northwest states have made limited efforts to encourage efficient water use, they have been very deferential to current water uses and practices. None of the four state water resource agencies in the Northwest believes it has adequate resources to do its job. When asked by agency headquarters for an estimate of additional staffing needs, one OWRD field manager responded as follows: You asked for a "Cadillac" and a "Citron" [sic] version of what staffing additions are needed to regulate to rate and duty on all Snake River tributaries. Our needs estimate is more of what you might call a "Chevy" or "Dodge" version, like the earlier Grande Ronde proposal. This "Chevy" version is believed to allow us to regulate to rate and duty in two years.

We think that this is reasonable and justifiable. The minimum needs estimate is just that, a minimum. This staffing level will not get us to rate and duty, but is hoped to be sufficient to let us identify problem areas. Not much more than [the watermaster's] description of our current program, "a one speed bicycle."

Despite increasing pressures on water resources in the 1990s, none of the state legislatures have provided any significant increases in agency enforcement resources. The states are not the only water managers in the Northwest with a poor record of enforcing legal requirements; the U.S. Bureau of Reclamation (Bureau), too, has been passive in ensuring compliance with federal laws and contracts. The Bureau has known for well over a decade of widespread misuse, or unauthorised use, of water from federal reclamation projects. This "water spreading" arose in part because the Bureau had consistently failed to ensure that project water was being delivered in accordance with legal and contractual requirements. In 1994, the Bureau promised that it would act to resolve water spreading, but soon its officials were assuring project irrigators that it had taken the wrong approach. Shortly thereafter, the Bureau quietly abandoned the issue over the protests of environmentalists.

The agencies' failure to enforce is somewhat understandable, in part because any efforts they make to curtail existing water uses may be vitiated by the legislatures or courts. Consider the experience of OWRD in its attempts to regulate certain groundwater users whose pumping affected surface flows in the Umatilla River Two water users filed suit, and the district court for Umatilla County blocked the Department from regulating them in conjunction with surface water. In addition, the 1995 Oregon legislature approved a bill intended to block such regulation, although the governor vetoed it.

Change the Law Where Needed to Protect Existing Uses

Central to the status quo policy is the notion that existing laws are generally less important than existing water uses. As explained in the following paragraphs, the Northwest states in the 1990s enacted numerous revisions to their water laws that help preserve established water uses. In addition, the 1997 Washington legislature approved several bills that would have significantly altered state water law in favour of status quo water uses, although Governor Locke vetoed all or part of ten such bills.

The Idaho Supreme Court's 1994 decision of Musser v. Higginson created something of a crisis in the state. Under prior appropriation law, the Musser case was hardly revolutionary; the court unanimously held that the Idaho Department of Water Resources (IDWR) had a clear legal duty to distribute water in accordance with rights of prior appropriation. Musser threatened a serious upset of the status quo because it called on the state to curtail groundwater use from the Snake plain aquifer in favour of senior surface water appropriators. Consequently, IDWR quickly adopted conjunctive use rules that, while giving a nod toward prior appropriation, authorised (continued) "reasonable use" of both surface and groundwater resources, Thus, after the Idaho Supreme Court forced a reluctant IDWR to curtail junior groundwater users, IDWR established more "reasonable" rules, that is, rules more favourable to junior pumpers than the traditional tenets of prior appropriation.

The Idaho Legislature has also acted repeatedly to legitimize existing, otherwise illegal irrigation uses through the Snake River Basin Adjudication (SRBA). The legislature protected water users who had irrigated an excessive number of acres, or who had changed their place of use without the necessary state approval, Idaho enacted statutes in 1985 and 1989, attempting to grant such illegal uses a presumption of validity, but the meaning of these laws was exceptionally opaque, and the court in In re SRBA declared them unconstitutionally vague in 1994. Within weeks, the Idaho legislature passed new statutes that clarified and strengthened protection for otherwise illegal uses in the SRBA. The legislature declared that the public interest was served by confirming past expansions and transfers of water rights that had been made out of compliance with state law. The Idaho Supreme Court upheld the new statutes as constitutional.

Montana and Washington also changed state statutes to resurrect certain forfeited water rights. Montana law originally prescribed a deadline of June 30, 1983 to file claims of existing rights in the statewide general adjudication. Users filed a mind-boggling 216,000 claims, and the adjudication rumbled forward. Nevertheless, in 1993 the Montana legislature authorised the filing of "late claims," essentially granting amnesty to those who had lost their rights by failing to file ten years earlier. Washington took the same action three times, enacting statutes in 1979, 1985, and 1997 to allow the filing of precode water right claims, even though state law had provided that such claims were conclusively deemed waived and relinquished if not filed by 1974. Both states thus reinstated old water rights that had already been forfeited by operation of law.

The Oregon legislature acted specifically to preserve existing water use by the Grants Pass Irrigation District, and to save the district's embattled Savage Rapids Dam. Savage Rapids is an aging diversion dam on the Rogue River that poses a serious barrier to salmon passage, and also diverts water at a far greater rate than the district needs to irrigate its acreage. Oregon officials found that the district's diversions exceeded its water rights, and refused to issue a new right allowing the practice to continue. After years of conflict, the district struck a deal with the Oregon Water Resources Commission and dam opponents: the district would receive a temporary water right, allowing continued diversions at a somewhat excessive rate, while the district had to begin taking steps to remove the dam and replace it with pumps.

The Commission issued a permit on the basis of that agreement in 1994. But the following year, the Oregon legislature approved two bills in an effort to save Savage Rapids Dam. The first would have legislatively issued the district a new water right with no conditions regarding dam removal. Oregon Governor John Kitzhaber vetoed that bill, chiefly on the grounds that the legislature should not issue water rights. The second bill blocked dam removal temporarily while a new task force, established by the bill, studied the matter. Kitzhaber signed that bill into law, effectively derailing the water right/dam removal agreement.

The Oregon legislature first authorised "amnesty" for water users in 1989, establishing a process whereby irrigation districts could legitimize otherwise illegal changes in their water use, subject to certain conditions. By the end of 1995, Oregon had created several more exceptions to the appurtenancy requirement, particularly for irrigation districts. Among other things, Oregon established a streamlined process for the transfer of water rights within irrigation districts, including water rights already subject to forfeiture for five successive years of nonuse. A 1995 statute allowed districts to obtain a certificate of water right for irrigation of lands not covered by the underlying permit. The legislature did, however, preserve the "no injury" test for all these changes.

Avoid the Position of Having to Curtail Uses

Agencies, courts, and legislatures have recognised that they can often preserve the status quo quite effectively by limiting state involvement in water rights controversies. This approach not only allows existing uses to continue, but also saves agency resources and reduces state entanglement in potentially nasty matters. Limiting state involvement may not be good public policy, but it often makes sense in both bureaucratic and political terms by keeping agencies and elected officials from taking unpopular stands against established water uses.

The Washington Supreme Court's 1993 Sinking Creek decision, Rettkowski v. Department of Ecology, severely constrained Ecology's authority to administer water rights. In the Sinking Creek case, Ecology had stepped in (after about twenty years) to regulate junior groundwater irrigators in favour of senior surface water ranchers. A majority of the Washington court, however, held that Ecology had no statutory authority to regulate water use until a general stream adjudication had been conducted in the basin. Two justices entered a persuasive dissent on legal, practical, and policy grounds. Among the points made by the dissent:

> The majority correctly points out that its decision will not provide for a "cheap and easy" water adjudication solution. Prohibitively expensive and interminable litigation is what the majority has fashioned as a solution, and to no purpose. The relief sought by neither party was for a general adjudication, and yet that is now the only relief which the majority opines is available. The director of Ecology, upon reading the majority opinion, will surely scratch her head in wonderment that she has the responsibility for issuance of water use permits but no authority to regulate those permits.

In the years since the Sinking Creek decision, the Washington legislature has done nothing to restore Ecology's regulatory authority over water use. Thus, for the majority of Washington river basins not yet adjudicated, the state exercises no significant control over existing water uses.

Idaho is currently undergoing its own "expensive and interminable litigation" to adjudicate the water rights of the Snake River Basin, which covers nearly the entire state. A 1996 decision of the SRBA court, however, raised profound questions about whether the adjudication would actually scrutinize current water use in any meaningful way, or simply confirm "paper rights." SRBA-Judge Hurlbutt ruled that Idaho statutes did not allow for "partial forfeiture" of water rights.

In other words, forfeiture of a water right for nonuse would be an all or nothing proposition; unless completely forfeited, water rights would not shrink after obtaining their maximum size. The Idaho Supreme Court overturned this decision, holding that Idaho law does provide for partial forfeiture. Had Judge Hurlbutt's ruling stood, it would have resulted in SRBA decrees for water rights considerably in excess of current uses. At the very least, it would have reduced agency and court involvement in major disputes over forfeiture, which would not have become an issue unless a right had been entirely unused for several years.

In Montana's massive statewide adjudication, DNRC has scaled back its efforts to evaluate and oppose water use claims. The agency is less active and aggressive than it used to be, largely due to pressure from the water court itself. Such a policy may in fact speed up the adjudication, which has long been an interest in Montana, but it also protects existing users who generally have the most to lose from a thorough evaluation of their water rights. As one commentator has stated:

Water right adjudications promise secure rights, finality, and better management of water resources. But uncertainty and ineffectual state control favour the status quo ante users, most of whom are also in a position to control the pace of adjudications through litigation and the effectiveness of adjudications through state legislatures. Claimants in control of large blocks of water have only lukewarm enthusiasm for decrees and stridently challenge any decision or order that diminishes their claimed rights. Every set of water users faces the possibility that a final decree and better administration may limit or interfere with their accustomed water use. The State of Oregon, as a policy matter, has chosen not to regulate water use in the Klamath Basin until the Klamath adjudication is complete. At the same time, Oregon has essentially taken the position that the U.S. Bureau of Reclamation, which owns and manages the Klamath reclamation project, may not deliver water for any purpose but irrigation until the adjudication is complete.

According to the state, the Bureau lacks authority to deliver water for fish, tribal needs, or wildlife refuges in the Klamath Basin, because the project's water rights are exclusively for irrigation. Thus, the state claimed that it had the sole authority to allocate water and regulate its use, but declined to exercise that authority. Oregon's stance may be legally suspect, but it is politically and bureaucratically expedient. The state's position is at once pro-existing user and anti-federal government, and while asserting state authority, it actually limits state involvement. Some observers believe that watershed councils and other locally driven, consensus-based efforts have gained popularity in the Northwest largely because they can help guard the status quo. Whatever their potential benefits for resource protection and restoration, local consensus efforts seem likely to protect existing water uses, limit state involvement, and reduce controversy on contentious issues.

The trend toward local empowerment runs counter to traditional western water law, however, in that it tends to "localise" a resource that is legally controlled by the state and owned by the public.

Prevent Instream Demands from Threatening Existing Uses

Water users often voice support for instream flows, but user groups typically do not want to see streamflows protected and restored at their expense. Thus, the status quo policy moves states to ensure that instream flow demands do not threaten existing uses, either through some specialised legal approach or under the traditional rules. This aspect of the status quo policy is perhaps best illustrated by the basic character of state instream flow programs in the Northwest. The laws of Idaho, Oregon, and Washington all essentially limit the establishment of minimum protected streamflows to state agencies, while Montana's law is just slightly less restrictive. These arrangements effectively allow politically powerful user groups, with their influence in state government, to maintain considerable control over instream flow protections. These programs also relegate instream rights to second-class status in a system that one commentator has deemed "socialist":

> Barriers still stand to the full integration of instream uses into state water allocation systems. Barriers (such as the "second class" treatment of instream flows) are not due to any natural incompatibility between instream rights and the prior appropriation doctrine. Far from it; the doctrine is fully up to the task of accommodating modern needs. All that remains is to eliminate those legislative restrictions which are essentially alien to the doctrine's nature.
>
> The most pervasive and frustrating limitation on instream water rights in most western states is the prohibition against ownership of instream rights by persons other than a designated state entity. This prohibition is a curious twist on the prior appropriation doctrine. It reflects a basic discomfort with the concept of instream rights and an underlying distrust of those entities which may seek to acquire them, particularly environmental groups and the federal government.
>
> Ironically, then, the drafters of such programs have turned away from the market-based principles which underlie the prior appropriation system, and have embraced principles of command-and-control resource allocation socialism, if you will with respect to instream rights.

Conceptual problems aside, the Northwest states' instream flow programs face many practical obstacles. These programs must overcome procedural burdens, inadequate enforcement, minimal funding, and political pressure in order to protect streamflows in any meaningful way. These limitations undoubtedly reflect existing users' discomfort with instream flow rights, even where those rights would be later in time and thus lower in priority. Washington Governor John Locke vetoed portions of a bill passed by the 1997 Washington Legislature that would have largely transferred instream flow protections into local hands. The bill would have established local "planning units," with multiple votes for local governments and water users, but only one vote for the Washington state government. These planning units would have had the authority not only to determine future instream flows, but to "adjust" streamflow protections already established by the state. But for the Locke veto, all of

Washington's existing minimum flows could have been eliminated, watershed by watershed, by vote of local governments, water users, and property owners. The vetoed legislation would have shifted to the local level a large measure of the state's traditional control over a public resource, although the legislation would not have allowed local planning units to "adjust" existing out-of-stream water rights. Private transactions to convert existing water rights to instream use, whether permanently or temporarily, could provide some senior water for flow restoration. Each of the Northwest states makes some provision for such transactions, while maintaining significant limitations.

The 1995 Montana legislature authorised a time-limited program allowing temporary transfers of water for instream flow use; a cautious step, certainly, but still a step in the direction of expanding options for flow protection. In Oregon, by contrast, bills were introduced in the 1995 and 1997 legislative sessions to prohibit any transfer of water from irrigation to any other use. The U.S. Bureau of Reclamation supported changing Idaho law to allow the transfer of natural flow water rights for instream purposes, but it could not even get the bill introduced in the 1997 Idaho legislature.

Environmentalists sometimes suggest eliminating wasteful water use as a promising method for restoring depleted streamflows. The Washington Supreme Court recognised that possibility in State Department of Ecology v. Grimes, but seemed to foreclose it. In Grimes, the court reviewed a referee's report that seemed to consider environmental factors in evaluating the Grimes' claims in a water rights adjudication. The court noted that, while considering environmental impacts is consistent with certain state obligations under the water laws, "these factors cannot operate to impair existing water rights." In fact, the court specifically rejected the use of a "reasonable efficiency" test for reviewing existing water rights, saying it was inappropriate in the context of an adjudication.

The Idaho Supreme Court similarly rejected the application of the public trust doctrine in the Snake River Basin Adjudication (SRBA). The court confirmed that water rights in Idaho are held subject to the public trust, which can take precedence over vested water rights. But according to the court, the public trust was not an element of a water right, and therefore not an issue to be determined in the SRBA. The Idaho legislature acted quickly to address a perceived threat to established water uses by passing a statute in 1996 that purports to preclude application of the public trust doctrine to water rights.

Effects of the Status Quo Policy

The basic rules of western water law work to the benefit of some groups of people and the detriment of others. Whatever else might be said about them, the traditional rules at least give all interest groups a clear sense of what to expect if the rules were in fact applied. Senior users would have a reliable water supply, but would be limited to the terms and conditions of their water rights. Junior users' diversions would be curtailed, if needed, to satisfy senior rights, but senior rights could not be changed or expanded to the juniors' detriment.

Instream flows would have very late priorities, but they would nonetheless be protected to that extent. They would benefit over time as rules, such as waste and abandonment, were applied against consumptive rights. Prospective new users would be unable to obtain a new water right from a fully appropriated source. However, they would have an opportunity to acquire an existing right and change its use, perhaps transferring the water to a new type of consumptive use or even instream use. But under the status quo policy, the results are often quite different from these expectations.

Winners and Losers Change

When compared to what would happen if the traditional water law rules were applied, if a state preserves the water use status quo, it often helps one class of existing water users to the detriment of another. A legislature, court, or agency may understand that it is helping one interest at another's expense, but in some instances may not realise that it is effectively redistributing water from one group to another.

Many decisions essentially have involved a choice among irrigators: surface water users versus groundwater users, juniors versus seniors. Between these classes, the winners have not been consistent, but the states have consistently preserved the status quo.

In other words, the states have chosen to allow water users to continue their established practices, rather than apply the traditional rules of prior appropriation in a manner that would change those practices. The Oregon and Idaho legislatures effectively chose seniors over juniors (who would have benefited from the enforcement of water laws and water rights) when they offered a sort of amnesty to many existing water uses.

On the other hand, the Idaho Department of Water Resources's conjunctive use rules, issued in response to the Musser decision, effectively favoured junior users who pumped groundwater over seniors using surface water. The Washington Supreme Court maintained the status quo, thereby helping junior irrigators at the expense of seniors, when it took a narrow view of the Department of Ecology's authority in the Sinking Creek case.

In these latter two instances, both seniors and juniors had long used water with little state involvement. The seniors would have benefited from curtailment of the juniors' pumping in times of shortage, but the Idaho rules and the Sinking Creek decision created obstacles to state regulatory action.

The Sinking Creek decision illustrates how the status quo policy can harm even senior water users who have tried to defend their rights. The Sinking Creek ranchers held senior rights to surface water, and, although they began complaining to Ecology in the mid-1960s that they were being harmed by junior irrigators' groundwater withdrawals, the agency took no action until 1989. When the agency did take action, the junior pumpers sued to block regulation, and the Supreme Court ultimately held that Ecology lacked statutory authority to act. Thus, as the Sinking Creek dissenters noted, the seniors were left with no good options to settle their long-running dispute. The dissenters openly urged the Washington legislature to fill the management void created by the majority opinion, but so far the legislature has not done so.

Protection of established water uses may also harm prospective new users. By refusing to apply laws that would reduce the quantity of water legally held by existing users, states limit the amount of water potentially available for new appropriations. For example, Montana and Washington could have freed up some water for new uses by applying their forfeiture laws against users who had failed to submit timely claims, but the states chose instead to protect the status quo by allowing the filing of late claims. Even though improving efficiency could provide water for new uses, none of the Northwestern states has taken serious action against waste.

States also may disadvantage prospective new uses by restricting transfers of existing water rights. For example, Montana prohibits the transfer of large water rights unless the applicant can prove that several conditions are satisfied. Oregon constrains interbasin water transfers to protect the basin of origin. Idaho law prohibits changing the purpose of use of agricultural rights if the change "would significantly affect the agricultural base of the local area." Bills introduced in the past; two sessions of the Oregon legislature would prohibit any water transfers from irrigation to any other use.

Instream uses probably suffer most under the status quo policy. Like prospective new consumptive uses, instream flows generally lose out when states refuse to require existing uses to comply with established laws and water right conditions. Flow restoration efforts are also undermined by laws and attitudes restricting water right transfers and leases for instream use. For example, Idaho law precludes transfer of consumptive water rights for instream use, even though the state water plan calls for that prohibition to be changed. Moreover, states sometimes disarm instream flow advocates by preventing them from employing certain legal tools. For example, the Idaho legislature effectively shut conservation groups out of the Snake River Basin Adjudication. Similarly, the Washington Supreme Court in Grimes held that consideration of environmental values could not "operate to impair existing water rights."

Even in a crisis atmosphere arising from the prospects of salmon run extinction and increased federal regulation, the Northwestern states seem unlikely to take strong enforcement action against existing users in order to protect streamflows. A recent "crisis" in Oregon arose from the proposed listing of two stocks of coho salmon under the federal Endangered Species Act. State political leaders pushed hard against a listing, urging the federal government to defer to Oregon's own plan, the Oregon Coastal Salmon Restoration Initiative. The state initiative contained several measures relating to water quantity and instream flows, addressing new instream and consumptive water rights, enforcement of existing rights and laws, increased water use efficiency, and other matters. Oregon's plan did not commit the state to taking aggressive measures on existing uses.

For example, the initiative stopped short of requiring water users to measure and reports their withdrawals, promising only that the state would consider it. The initiative called for four additional Water Resources Department field employees, but did not commit the Department to taking controversial enforcement actions such as limiting water users to their annual duty or regulating groundwater in conjunction with surface water. The state agreed to convene "collaborative" local groups to develop water use efficiency standards, but was vague as to whether it would establish standards in the absence of local consensus, or how it would use them in any event.

Continued protection of existing user groups will likely prevent meaningful progress toward restoring streamflows in most of the Northwest. The water use status quo means seriously depleted (even nonexistent) flows in many rivers and streams. New or recently established streamflow protections, at best, will merely prevent the situation from getting worse. Water right acquisitions for instream flows and public funding for water conservation offer win-win solutions that seem likely to work far better in theory than in practice, given the massive amounts of money required to restore flows in even one river. Economic and environmental costs will also limit new storage projects and other "structural solutions" to flow restoration. Cooperative local efforts based on consensus, for all their promise and recent popularity, offer little hope of addressing instream flow problems effectively. The conclusion is unpleasant but inescapable: protecting the water use status quo means accepting the continued dewatering of Northwest rivers and streams for the foreseeable future.

In general, existing water users in the Northwest have been remarkably successful at continuing their established practices even where these practices violate venerable principles of western water law, and even though applying that law would often benefit other interests. The reason for existing users' influence in maintaining the status quo is open to debate, but public choice theory offers an interesting insight. One branch of public choice theory suggests that groups representing narrow economic interests are often remarkably effective in securing favourable results from public bodies. Research on interest groups' influence in Congress shows that a group is most likely to succeed when the group 1) is attempting to block, rather than obtain legislation; 2) has goals that are narrow, with little public visibility; 3) is supported by other groups and public officials; and 4) can shift the issue to a favourable forum, such as a friendly congressional committee.

This research suggests some of the reasons why existing users often win; they seek to preserve the status quo (although they sometimes have done so by having legislation passed, rather than blocking it), they have limited goals in low-visibility issues, they typically have strong support from their legislators and other political figures, and they often have favourable venues in legislative water committees and local courts, if not in the state agencies themselves.

Economic Questions Persist

The state and federal governments commonly justify the protection of the water use status quo on economic grounds. Thus, Idaho offered "amnesty" to otherwise vulnerable uses in the Snake River Basin Adjudication to protect "significant investments by water users and tax base for local governments." Similarly, the U.S. Department of the Interior has been reluctant to attack subsidies or even unauthorised water uses, because users have come to depend on them. The rationale appears to be that individual livelihoods and local communities have come to rely on existing water use arrangements and it would be unfair or unduly costly to disrupt them, regardless of the problems they pose.

Certainly there are some people who have come to depend (directly and indirectly) on existing water uses, and their lives would be disrupted if those uses were discontinued. In other contexts, however, this rationale is not enough to justify continued violation of the law, or especially continued expenditure of public funds. The Umatilla Tribes have noted the disparity in treatment between illegal tribal fishing and illegal irrigation: "When Indians fish illegally, we are sent to federal prison. When irrigators kill fish by illegally taking water, they are not punished. Instead, we are told by the United States Government that we must consider the impacts to the irrigation economy of making them stop their illegal activity." Military bases, too, have closed despite their economic importance to local communities. The fairness of protecting existing uses based on economic reliance is arguable, depending on one's point of view.

Even from a purely economic standpoint, the status quo policy seems unsound in many cases. On one hand, existing water uses certainly provide economic benefits to their users, their local communities, and the Northwest states. User groups and their allies emphasize these benefits in defending the status quo. On the other hand, existing uses impose various kinds of costs on society, both by causing harm to valued resources (such as dewatering salmon habitat or degrading water quality) and by preventing water from being used for other purposes (such as hydropower generation, new businesses, or instream flows). A comprehensive and accurate look at the costs and benefits of water uses may show that many current uses are far from optimal.

Critics of the status quo point to a variety of natural resource problems associated with current water uses. Such problems carry economic costs, although these costs are hard to quantify because of difficulties in valuing ecosystem services and other benefits of resources "in place." Clear identification and valuation of such costs and benefits would force the Northwest to reassess the economic worth of many current water uses. The public value of some uses may be far lower than commonly believed. Even economically beneficial uses, for example, irrigation of high-value crops in Idaho, may prove suboptimal at the margin, because society might benefit if some portion of the water used for irrigation went to other purposes instead.

Critics also maintain that, compared with alternative water uses, existing uses often generate little economic return per unit of water. Irrigation of alfalfa and other forage crops is most easily questioned, because these crops have relatively low values and high water demands. The marginal value of irrigation water in the West is commonly estimated to be less than fifty dollars per acre-foot. Meanwhile, municipal water suppliers have been willing to pay upwards of $2000 per acre-foot in recent transactions across the West. In some places, the highest economic use of water may be instream. From a purely economic point of view, perpetuating lower valued water uses is not efficient. As one commentator has stated, "although individual users, whether in cities or in rural areas, are probably efficient from their point of view, at the system level the vast differences in the relative value of water and often neglected instream values suggest that inefficiencies abound." Facilitating transfers from current uses to new uses of higher value, including instream uses, could produce real economic and environmental benefits.

Here again, public choice theory recognises that special interest groups may obtain outcomes that are not economically efficient from society's point of view. As scholars point out, this is not necessarily a bad result, because economic analysis is an imperfect tool for evaluating public policy questions of fairness and other issues must also be considered. But if the status quo policy produces economically inefficient results, decision-makers should openly and clearly explain why protecting existing users is in society's best interests. One interesting aspect of the status quo policy is that it sometimes contradicts that commonly stated maxim: "water flows uphill toward money." While that maxim seems likely to hold true over the long run, the current trend is to preserve existing uses. One commentator has observed that recent attempts to block out-of-stream or basin transfers amount to "quasi-riparianism." Money can sometimes make water move uphill, but existing users often can stop water from moving anywhere. Thus, existing water users may be even more influential than money.

Implications of the Status Quo Policy

Despite the pressures of population growth, environmental restoration, and economic transformation, established water uses have remained solidly protected in the Northwest. In order to perpetuate current uses, state legislatures, courts, and agencies alike have refused to apply, and sometimes have even changed, legal requirements. The record clearly shows that the states have been extremely reluctant to implement the traditional rules of western water law if doing so would mean upsetting the water use status quo. Today, those traditional rules are often honoured in the breach; if they ever were sacred, certainly they are not any longer.

In managing water resources, the Northwest states rarely deviate from the status-quo policy. Like it or not, anyone with a stake in the resource should recognise that the states generally will do what they must in order to protect existing water uses. Legal rules, agency mission statements, and even the interests of other water users usually are not enough to force action that would threaten established uses. Given this reality, stakeholders and decision-makers should approach water-resource issues somewhat differently than they would under the traditional rules. No one, not even the owner of vested senior water rights, should assume that her interests are safe simply because she is entitled to protection under basic water law rules. Those rules routinely have not been applied or have been changed, harming junior appropriators sometimes, seniors sometimes, and instream flows nearly every time.

Anyone with an interest in a water-resource must actively and vigilantly defend it, or risk having that interest impaired for the sake of some other user group. In addition, anyone considering a proposal, strategy, or scenario involving water use should evaluate it on the assumption that the status quo will be protected. For example, in evaluating whether a state management plan that emphasizes water law enforcement is sufficient to avert an Endangered Species Act listing, the federal government should ask: Is this likely to work? Will the state actively enforce the law against existing water users and make it stick? A group of irrigators considering a lawsuit to protect their water supply against other users should ask: Can I expect a court to shut off existing users? If so, will the legislature step in to protect them?

A state agency establishing policy for water use during droughts should ask: If we waive certain rules for water users during drought periods, will we be able to reinstate those rules when the drought is over? Or will the users find a way to continue the practices they started during the drought? Obviously, it would be foolish to assume that the law will never mean anything at all. But one should question any approach that relies on any level or any branch of government to enforce the law against established water uses.

Finally, while the states obviously have great concern with protecting existing water uses, they otherwise seem to have little regard for preserving western water law. The elements of traditional water law that remain most strong and vital water rights as property rights, which last forever unless abandoned, with priority based on seniority are the elements that most effectively safeguard established uses.

Most of the other elements are either moribund (such as the prohibition against waste), rarely enforced (forfeiture), or readily changed (appurtenancy), To the extent that the traditional rules remain relevant, it is not because they provide a legal framework of any inherent rightness or power. Instead, they persist largely because water politics in the 1990s still favours existing water users who defend their current practices and continuing entitlements based on economics and equity.

Traditional western water law does not directly recognise political, economic, or equitable concerns. But by managing water based largely on these factors, states tacitly recognise that prior appropriation forms an imperfect and incomplete basis for water policy and management. In a sense, when states go outside the traditional rules of western water law to preserve the status quo, they sacrifice prior appropriation for a kind of "public interest," even where the interest being protected is a private water use that actually violates current laws.

In sum, by consistently choosing to protect established water uses rather than applying the familiar rules of prior appropriation, the Northwest states have significantly undermined those rules. The region's water law and policy is now based more on today's notions of politics, economics, and equity than on any set of long-standing, bedrock principles. These notions will eventually change, however, and the water users now benefiting from the status quo policy may someday wish that the states had been more faithful to the traditional rules of western water law.

2

New Generation of Indicators for Coastal Waters

More than half the world's human population lives within 100 km of the coast, and that number is expected to increase by 25% over the next two decades. Consequently, coastal ecosystems are at serious risk. Larger coastal populations and increasing development have led to increased loading of toxic substances, nutrients and pathogens with subsequent algal blooms, hypoxia, beach closures, and damage to coastal fisheries. Recent climate change has led to the rise in sea level with loss of coastal wetlands and saltwater intrusion into coastal aquifers.

Coastal resources have traditionally been monitored on a stressor-by-stressor basis such as for nutrient loading or dissolved oxygen. To fully measure the complexities of coastal systems, we must develop a new set of ecologic indicators that span the realm of biological organisation from genetic markers to entire ecosystems and are broadly applicable across geographic regions while integrating stressor types. We briefly review recent developments in ecologic indicators and emphasize the need for improvements in understanding of stress-response relationships, contributions of multiple stressors, assessments over different spatial and temporal scales, and reference conditions.

We provide two examples of ecologic indicators that can improve our understanding of these inherent problems: a) the use of photopigments as indicators of the interactive effects of nutrients and hydrology, and b) biological community approaches that use multiple taxa to detect effects on ecosystem structure and function. These indicators are essential to measure the condition of coastal resources, to diagnose stressors, to communicate change to the public, and ultimately to protect human health and the quality of the coastal environment.

More than half the world's human population resides within 100 km of the coastline [National Oceanic and Atmospheric Administration], with increases likely over the next two decades. The coastal zone represents at least half the value of global ecologic services, and in economic terms, is the single most important source of recreational and residential income worldwide as well as fisheries.

Human development of coastal watersheds has greatly accelerated environmental pressure on downstream estuarine and coastal ecosystems; yet, unfortunately, assessing detrimental changes in these systems is complex. The symptoms of degradation include deterioration of water quality, loss of habitat and biodiversity, beach closings, fishery declines, fish consumption advisories, and an overall decline in the livability in the coastal zone. Coastal systems are hydrologically complex and are among the most susceptible to direct disturbance through global climate change. For instance, sea levels have been rising over the past century and even greater rises are predicted over the next 50 years.

These changes will continue to affect coastlines and will dramatically increase saltwater intrusion into freshwater coastal aquifers as well as the displacement of coastal agriculture. Climate change is also believed to have increased events of heavy precipitation and flooding, which recently have become more common. These events increase the flushing of nutrients and toxic chemicals into coastal regions. In addition, many estuaries are highly urbanised and reference conditions are difficult or impossible to specify because of large changes in habitat, water diversion, and the introduction of exotic species.

Not only are estuarine systems complex, but many stressors are also difficult to control. For example, non-point source nutrient pollution from coastal watersheds is a major problem, that has affected more than 60% of coastal rivers and bays. Increases in nitrogen and phosphorus coming into coastal ecosystems have led to disruptions of basic ecologic functions [e.g., rising frequency and proliferation of harmful algal blooms (HABs) and increasing oxygen depletion (hypoxia)] with major damage to coastal fisheries and biodiversity. In addition, nonpoint sources of toxic substances (e.g., agricultural chemicals and urban runoff) have impaired habitat quality for aquatic life and the human use of numerous coastal watersheds.

It is increasingly evident that ecosystem and human health are intricately linked. For example, HABs can cause diseases in humans that result from consumption of contaminated seafood or inhalation of toxins entrapped in sea spray. Moreover, the distribution and frequency of HAB events have increased along U.S. coastlines over the last 30 years. People are also exposed to waterborne diseases through recreational contact, and the incidence of these diseases is increasing worldwide, as is the cost to the economy of frequent beach closures. The principal agents of these diseases are bacteria, viruses, and protists. In the case of bacteria, this includes both native marine organisms (e.g., Vibrio species) and human or animal-derived pathogens from sewage and runoff. The importance of these connections between coastal conditions and human health was realised by the National Institute of Environmental Health Sciences and National Science Foundation (Washington, DC), and they subsequently established their Centres for Oceans and Human Health.

Given the importance of coastal systems and the increasing pressure on them, quantitative measures of coastal ecologic conditions are absolutely essential for detection of change as well as for the design of control measures and restoration activities. Many approaches to comprehensive assessment of condition are cost prohibitive; thus, there has been a tendency to use broadly applicable indicators such as water clarity, nutrient or contaminant loads and levels, and various biodiversity measures such as species richness as metrics. Yet measurements of environmental condition are becoming more sophisticated and more applicable across space, time, and biological organisation.

These new indicators include such metrics as diagnostic photopigments of algal functional groups to assess eutrophication; biochemical and genetic indicators of toxicant exposure and stress; molecular techniques to assess human fecal bacterial distribution; isotopic techniques to evaluate nutrient enrichment; indices of biological integrity or other biological community responses; ecosystem and population modelling approaches; landscape metrics; and remote sensing techniques to detect large-scale land use impacts and change. While these indicators represent impressive advancements in both science and technology, there are limitations on their widespread and integrated use (NRC 2000).

Our primary goal in this review is to identify four critical areas in which scientific advancements are needed before improvements can be made in indicator development of coastal regions. In addition, we provide examples of two promising approaches to improvements in indicators. Defining the limitations of previous approaches and developing new approaches was a major goal of the U.S. Environmental Protection Agency (U.S. EPA) Science to Achieve Results Program (STAR) in establishing the Estuarine and Great Lakes (EaGLe) Research Program. The EaGLe investigators (Atlantic, Pacific, Gulf of Mexico, and the Great Lakes coastal areas) realise there is increasing urgency to develop indicators capable of detecting and diagnosing environmental conditions over space and time at cellular, organism, habitat, ecosystem, and regional levels. As the explosion of technical and conceptual advances in various disciplines ranging from molecular biology to ecosystems ecology and from remote sensing to bioinformatics continues to provide new and better tools, the goals of the EaGLe program and the science surrounding them are works in progress.

Limitations of Current Coastal Indicators

Most indicators were designed to provide specific information on local conditions such as water clarity or eutrophication or to provide broad-based snapshots of regional-scale water quality and habitat condition as exemplified by the total maximum daily load or Environmental Monitoring and Assessment Program (EMAP). From these studies we know that large areas of the U.S. coastal zones are impaired, and many of the specific stressors within coastal watersheds that contribute to impairments have been identified. Traditionally, the focus has been on indicators associated with water quality and toxic substances; however, there are major concerns on many other stressors, including habitat and landscape change, exotic species, global climate change, and pathogens. We focus on some of the limitations of current indicators that must be overcome for future advancements to be made on broad aspects of this problem in the coastal region.

Stress with response. Most current indicators of coastal condition are not linked with specific stressors; hence, it is unclear what causes the change in the indicator and, therefore, what management solutions should be implemented to affect an improvement.

Multiple stressors. Stressors in coastal ecosystems are diverse and originate from both anthropogenic and natural perturbations. Most current indicators are incapable of providing diagnostic information and are unable to discern the relative contribution of the various stressors to the observed changes in the indicators. Space and time. Sources of stress to coastal environments operate over a range of spatial scales (e.g., square metres to entire landscapes) and time (seconds to decades). Current indicators are typically not explicit in how they relate condition with stressors over these different scales.

Reference conditions. The interpretation of the condition or change of an indicator is based on a comparison to reference conditions or benchmarks. Frequently, these reference conditions are not quantitatively defined; thus condition or meaning of a change in indicators is subject to considerable interpretation and debate.

Linking Stress with Response

From a management perspective, among the greatest limitations of many indicators of coastal condition is the lack of a link with the cause for change.

In the development of indicators, distinguishing between measurements of disturbance or stress and measurements of ecosystem response is imperative. The terminology in this area of the literature varies considerably, but recent reviews provide clarification.

In developing stress response relationships, natural experiments or surveys across a disturbance gradient from relatively pristine to highly disturbed areas are used most often. In these situations, it is often difficult to control the stress because multiple factors often vary across any environmental gradient. Thus, inherent limitations in the scope of inference result and must be explicitly recognised. Understanding the response variable is also dependent on the strength of conceptual models used to describe the factors structuring the ecosystem, and on the extent to which anthropogenic disturbances influence that ecosystem. Obviously, detection of a response to a stressor is best accomplished by experiments in which the stressor can be manipulated in frequency, intensity, and duration. Unfortunately, these experiments often lack realism, as they are typically limited to laboratory situations or small-scale field-based mesocosms.

However, a combination of a gradient design with field and laboratory experiments can be a powerful approach for the initial phases of indicator development, but as the indicator is applied at larger spatial scales or in uncharacterised sites, additional approaches are needed. Multivariate statistical approaches, novel modelling approaches, and techniques to aggregate indicators according to their use in management may all be valuable approaches. For example, these may permit more realistic diagnosis of stressors as complex as nutrient inputs, pathogen effects, and toxicant bioaccumulation. A number of existing approaches are available to couple toxicant stress with response. For example, evaluation of the effects of toxic substances on ecosystems can involve multiple approaches: a) comparison of a toxic response or specific dose level of a contaminant to a water quality standard that has been linked to biological effects; b) assessment of environmental effects of multiple toxic substances by using well-characterised individual contaminant exposures; or c) the coupling of physiologic and genetic indicators with environmental chemistry and ecologic responses at multiple spatial and temporal scales.

However, these too have their limitations. The first approach is limited when multiple contaminants and multiple stressor types are present. The second approach is primarily limited because often no direct integration of the toxic response and exposure occurs. The result is data correlation, which is also limited to the spatial and temporal scopes of the immediate investigation. The third approach considers multiple stressors and permits direct integration and scaling, but significant challenges remain to fully develop indicators for a range of habitats and model organisms. There are also combinations of these approaches. A simple index of sediment contamination, the Sediment Quality Triad, combines the first two. It provides a framework for analysing benthic community data, analytical chemistry, and toxicity test data to assess whether a site is affected by toxicants, and is widely used throughout the nation. Yet, the action levels derived for specific contaminants are often unknown and benthic data are highly variable. Moreover, there is often a lack of specific reference conditions that precludes dear interpretation. The triad approach is also based on acute lethality, whereas sublethal effects can also be very important.

In summary, there is a need for controlled experiments in laboratory and field settings for those stressors (e.g., toxicants, nutrients, turbidity) amenable to manipulation. For larger-scale stressors such as introduction of exotic species, climate change, habitat loss, or landscape change such as fragmentation that are not easily amenable to manipulation, field experimental designs that test responses over gradients of stressor levels are among the options for linking stress with response (Danz et al., in press). Coupling the approaches of laboratory and field methods are essential for future development of appropriate response indicators.

Multiple Stressors of Environmental Condition

Stress on coastal ecosystems is usually a combined effect of natural and anthropogenic disturbances. Natural disturbances in the U.S. coastal zones include water-level fluctuations from droughts and floods, wind events such as hurricanes, natural soil/sediment deposition, insect infestations, and forest fires. These natural disturbances vary in intensity both spatially and temporally. Major anthropogenic disturbances to the watershed of coastal ecosystems include permanent land cover conversions of native vegetation to agricultural, residential, and industrial areas; and temporary conversions of land due to forestry.

These conversions result in concomitant disturbance to coastal ecosystems, including a) landscape effects of fragmentation, b) increased surface water runoff, c) increased nutrient and sediment input, d) increased pesticide and other chemical inputs, e) increased water temperature, and f) greater human disturbance from recreational use, increased fish and shellfish extraction, and noise. Climate change and the resulting change in weather patterns is a combination of both natural, stochastic events, and human-induced warming which affects vegetation, water levels, and virtually all types of disturbance. In arid regions, water diversion and water use patterns also result in landscape and ecosystem-level alterations. Detecting the effects of both individual disturbances and the simultaneous influences of natural and anthropogenic disturbances in coastal ecosystems is a challenging and complex task.

The relative effects of anthropogenic disturbance must be distinguished from the ranges of variation in natural disturbance regimes, but because of the large size and variability of coastal ecosystems, manipulative experiments to untangle the complexities of the varying disturbance regimes are difficult except on a relatively small scale. Combining specific indicators with modelling efforts can clarify and distinguish anthropogenic from natural stress in individual ecosystems and regions.

However, the general trend has moved away from using direct diagnostic measures of stressors to using integrated indicators of ecosystem structure and function. Yet, we know this approach may be inadequate for many applications because stressors vary among regions, implying that indicators are needed with diagnostic and prognostic capabilities. Indicators, therefore, could be grouped to fit regional needs related not only to assessing condition but also to developing appropriate management responses.

Characterising the effects of multiple stressors on any ecosystem is among the most challenging tasks facing scientists today because multiple stressors can have synergistic, additive, or antagonistic effects on biological responses. Disentangling the various effects of multiple stressors will likely require a combination of controlled laboratory experiments, large-scale studies over multidimensional gradients of stress, and insightful modelling of ecosystem responses and change.

Spatial and Temporal Explicitness

Ecologic indicators are constructed or selected to assess the condition of ecosystems and to detect environmental change related to human disturbance. Condition is often assessed by documenting the state or rate of ecologic processes such as productivity, respiration, or the structuring of biological communities. Indicators may do this by either measuring those processes directly (such as primary productivity) or inferring process from pattern (such as indices of biotic integrity (IBIs) as descriptors of community structure). Ecologic processes operate over a range of spatial and temporal scales, and the resulting patterns are expressed over varied scales. Hence, the relevant scale of each indicator must be specified to relate pattern to process in the appropriate conceptual model. Levin (1992) stated "the concepts of scale and pattern are ineluctably intertwined. The description of pattern is the description of variation, and the quantification of variation requires the determination of scales."

Many studies have sought to quantify spatio temporal patterns across a range of scales. Unfortunately, few have determined whether patterns are consistent across scales or whether related phenomena cross scales. In addition, because of technical, logistical, and financial reasons, most ecologic studies have focused on small systems (e.g., site or plot level) and short periods of time, which in turn has limited the development over large spatial scales. Alternatives such as top-down approaches do exist. Here, large-scale processes form the basis of inferring process from pattern and are being applied in regional classification schemes.

The need for scale explicitness is complicated by multiple stressors arising from human and natural disturbances. Most ecologic indicators are related to multiple stressors and scales. For example, in a study of littoral macroinvertebrate communities, 23% of the variance in taxonomic composition was associated with habitat factors, but greater spatial scales (riparian, catchment, ecoregion classification) accounted for 24% of the variance. If indicators are to be used effectively in management, it is necessary that we know the relevant scale(s), so that the scale of management actions matches the scale of the phenomena being measured. Experimental approaches that allow for the partitioning of the variance among different stress components and over a hierarchy of spatial scales will be critical in the development of new ecologic indicators.

Reference Conditions

To interpret any set of indicators, one must compare results of monitoring to standards or benchmarks. One of the preferred benchmarks is a reference site or condition. The use of reference sites has become increasingly common as ecologists and managers search for reasonable and scientifically based methods to measure and describe the inherent variability in natural aquatic systems. As there are likely few places on earth unaffected by anthropogenic disturbances, true reference areas remain elusive.

For example, even coastal regions of Greenland or Antarctica have been affected by atmospheric chemical inputs and climate change. Alternatively, even in coastal regions with long histories of human occupation and, hence, anthropogenic disturbances, reference sites can be established using specific estuaries, watersheds, or lotic systems entering the coastal zone. These may represent the best attainable environmental conditions for a specific geographic setting, a historic representation using paleolimnologic data, a simulated reference condition, or a situation where conditions fall within the range of natural variability for the system.

Determining reference condition of a system is highly dependent on the indicators used and the locations where samples were gathered. Benthic indicators will provide different results than fish indicators. Similarly, indicators will be different in large, ephemerally stratified systems compared with smaller, well-flushed systems. For example, phytoplankton growth responses to nutrient enrichment will not be as profound as those for benthic microalgae in well-flushed systems. Here, benthic microalgae may be more sensitive and meaningful indicators of ecosystem response to nutrient enrichment. Indicators of community structure (i.e., diversity indices, keystone species) may gauge ecosystem conditions quite distinct from indicators of function (e.g., primary and secondary production, respiration, and nutrient cycling). IBI, habitat suitability indices, and chemical monitoring are specific examples of indicators that in combination can assess structure, physical chemical quality, and biological measures of reference condition.

Historical information from a site, such as survey data, paleolimnologic studies, and habitat reconstruction, can be extremely helpful for determining reference conditions. Finding such information, however, can also be very difficult and most historical information is not quantitative such as for urban estuaries. Hughes (1995) summarised the basic characteristics necessary for a suitable reference condition including reasonableness and political acceptability, sufficient number of reference sites within the area of interest, and suitable data on natural conditions of the site.

Examples of New Indicators

Environmental indicators can have an enormous number of possible end points, reflecting the breadth and diversity of the scientific underpinnings in biology, chemistry, and physics. For example, biological indicators span the realm of biological organisation from genetic markers to entire ecosystems. Chemical indicators reflect a variety of spatial or temporal scales ranging from oxygen demand for a specific point source to global carbon dioxide distributions in the atmosphere. Physical indicators can include elevational, morphologic, transport, circulation, exchange and stratification processes with all their attendant ramifications for ecosystem structure and function. Because of the massive amounts of information that can be gathered across levels of physical, chemical, and biological organisation and across spatial or temporal scales, the challenge to integrate data across levels of organisation in space and time is daunting.

Current programmatic and academic funding scenarios exacerbate our lack of integration. Most funding is limited by amount and duration. These monetary and time deficiencies have been recognised by academic funding sources, such as NSF and U.S. EPA (EaGLe). As such programs mature, advances in integrating across a variety of trophic levels as well as organisational or spatial and temporal scales will likely occur.

We provide two brief examples of new types of indicators; one that links productivity (function) and hydrology and another linking community (structural) patterns. We believe these new types of indicators will substantially improve our ability to measure and understand the complexity, response, and condition of coastal systems. For example, analyses of photopigments provide a means to explicitly link nutrient and hydrologic stressors with specific phytoplankton groups and over explicit spatial scales when combined with remote sensing information.

If data are gathered systematically over time, then temporal changes can also be linked with specific stress events (e.g., hurricanes). In the second example, multitaxa types of approaches provide a wide range of possibilities for improving our knowledge of stress response relationships, the identification of multiple stressor effects, spatial and temporal explicitness, and the identification of suitable reference conditions. For example, Luoma et al. (2001) provide eight case studies that specify cause and effect linkages using community analyses of macroinvertebrates to decipher the effects of multiple stressors. Paleolimnologic data derived from sediment and water column sampling of diatom communities are among the most powerful techniques for identifying suitable reference conditions within aquatic systems.

Photopigments as Integrators of Estuarine Nutrients and Hydrology

Nitrogen availability most frequently controls microalgal and higher plant primary production in estuarine and coastal waters. Loading rates of this nutrient directly reflects human population density and activity in coastal water and airsheds. Excessive nitrogen loading is a key causative agent for accelerating primary production or eutrophication. Symptoms include phytoplankton blooms, which may accumulate as ungrazed organic matter in the sediments, providing the "fuel" for oxygen consumption and depletion in bottom waters and sediments. This chain of events is particularly problematic in salinity- or temperature-stratified waters, where oxygen may not be easily replenished from the atmosphere. Hypoxic conditions alter nutrient cycling and promote fish disease and mortality.

Suspended microalgae or phytoplankton account for the bulk of estuarine and coastal primary production in many estuarine and coastal ecosystems. Their composition and activity are key in determining fertility, eutrophication, and water quality. Water discharge controls transport of phytoplankton through these systems and plays an interactive role with nutrient supply to control phytoplankton growth, competition, succession, and community composition. For example, high rates of freshwater discharge reduce the salinity and residence time.

These conditions favour fast-growing oligohaline phytoplankton, such as chlorophytes (green algae). In contrast, low-discharge conditions promote long water residence, high salinity conditions, which favour slower growing halophylic taxa such as dinoflagellates and certain cyanobacteria. Phytoplankton community composition affects the structure and function of estuarine food webs, nutrient cycling, habitat condition, fishery resources, and overall ecosystem condition.

Chlorophyll a has been used for many years as a sensitive indicator of phytoplankton biomass. However, because virtually all phytoplankton contain this pigment, it alone cannot be used to determine community composition. Using additional diagnostic chlorophyll and carotenoid photopigments as indicators of major phytoplankton functional groups (i.e., diatoms, dinoflagellates, chlorophytes, cyanobacteria, cryptomonads), we can examine the interactive effects of nutrient and hydrologically driven changes of phytoplankton community composition and activity. HPLC, coupled to photodiode array spectrophotometry is used to determine phytoplankton group composition based on the diagnostic photopigments. Photopigment markers include chlorophyll b and lutein (chlorophytes), zeaxanthin, myxoxanthophyll, and echinenone (cyanobacteria), fucoxanthin (diatoms), peridinin (dinoflagellates), and alloxanthin (cryptomonads). A statistical procedure, ChemTax, partitions chlorophyll a (i.e., total microalgal biomass) into the major algal groups to determine the relative and absolute contributions of each group.

Examples from ongoing studies in the Neuse River estuary in North Carolina and Pamlico Sound (1994-present) show that these systems have experienced the combined stresses of anthropogenic nutrient enrichment, droughts (reduced flushing combined with minimal nutrient inputs), and since 1996, elevated hurricane activity (high flushing accompanied by elevated nutrient inputs). Seasonal and hurricane-induced variations in river discharge, and the resulting changes in flushing rates, and hence, estuarine residence times, have differentially affected phytoplankton taxonomic groups as a function of their contrasting growth characteristics.

For example, the relative contribution of chlorophytes, cryptophytes, and diatoms to the total chlorophyll a pool was strongly controlled by periods of elevated river flow. These effects are due to the efficient growth rates and enhanced nutrient uptake rates of these groups. Cyanobacteria, on the other hand, showed greater relative biomass when flushing was minimal (i.e., longer residence times) during the summer.

Further evidence that hydrologic changes have altered phytoplankton community structure is provided by the observed historical trends in dinoflagellate and chlorophyte abundance. Both decreases in the occurrence of winter-spring dinoflagellate blooms and increases in the abundance of chlorophytes coincided with the increased frequency and magnitude of tropical storms and hurricanes since 1996.

The relatively slow growth rates of dinoflagellates may have led to their reduced abundance during these high river discharge events, These changes in the phytoplankton community have been linked to altered trophodynamics and nutrient cycling, which subsequently affects fishery habitats and yields. Diagnostic photopigment analyses are able to detect significant changes in phytoplankton community composition over a broad range of time scales (< 24 hr to decades) and as such are well suited for monitoring programs designed to assess short- and long-term trends in water quality in response to: hydrographic features (circulation, upwelling); nutrient enrichment; climatic; and hydrologic perturbations (floods, droughts). In addition "top down" effects of grazing have been examined using an HPLC-based technique. Finally, these analyses have proven useful as a means of ground-truthing and calibrating remotely sensed estimates of phytoplankton bloom events.

This coupling of indicator technology with remote sensing enables "scaling up," namely, mapping the spatial distributions of phytoplankton groups over large geographic areas not amenable to routine field sampling, evaluating the effectiveness of nutrient management strategies, use as an early warning system for blooms of nuisance or toxic species, and as a sensitive bioindicator of overall water quality conditions.

Biological Community Responses as Condition and Change Indicators

Plant and animal community structure and function have been extensively measured to describe the condition of both aquatic and terrestrial systems. The major challenge is how to scale and aggregate the responses of species populations at the site level to reflect conditions of the biological community level for specific taxa or to provide assessments of large scale patterns, such as IBI, biological species profiles, multitaxa indices, or indices of environmental integrity.

These approaches hold tremendous potential for assessments of environmental conditions over large landscape or regional areas, as well as for detection of temporal change. However, these indicators will also require considerable development in the areas of a) providing linkages with single and multiple stressors; b) exploration of analytical techniques to integrate and synthesize multiple biological signals from species or functional groups within the biological community; c) parsing these multivariate responses among stressors and over varying spatial scales; and d) providing explicit spatial or temporal scales for the indicators which are consistent with the scales of management actions.

The strength of the community approach lies in the differential sensitivity of individual species, functional groups (e.g., guilds), or trophic levels to different stressors. Each of these levels can respond differently to stress. For example, O'Connor et al. (2000) found a correlation among many taxa (diatoms, benthos, zooplankton, fish, birds) to the gross condition of lakes, but fish provided the best measure of condition in the near-shore environment. In an experiment on the pesticide effects of mosquito control agents in wetlands, of the zooplankton, aquatic insect, and bird communities studied, only the aquatic insects exhibited a response to treatment. In this case, aquatic insects were the best indicator of pesticide effects. Niemi and McDonald (in press) provide many examples of responses by different taxa to diverse stressors.

The unique aspect of the community approach is the ability to sample a wide variety of taxa; each of which has a unique life history capable of being disrupted by stress at various scales. All coastal regions are represented by thousands of species including taxa such as bacteria, plankton, macroinvertebrates, fish, vascular and nonvascular plants, amphibians, and birds. Many associations between these taxa and stress exist.

For example, diatoms are particularly sensitive to nutrients, whereas benthic invertebrates are responsive to sediment contamination in both lakes and estuaries. Fish communities are sensitive to human development and exotic species. Wetland vegetation is directly affected by hydrologic modifications such as dikes and road building. Amphibians are sensitive to water quality in wetlands. Bird populations are affected by landscape-level habitat change and fragmentation.

Moreover, many of these taxa have well-established sampling methods, and some have long-term nationwide monitoring programs that are currently still in use. The combination of species- or taxa-specific responses by plant and animal communities to stressors and the availability of extant monitoring programs allows for the partitioning of multiple stress-response relationships.

Probability-based and standardised sampling of communities within specific sites but over large landscapes has proven useful for regional-scale assessments of environmental conditions. For example, these approaches have been used for the identification of imperiled systems; development of biological indicators for the Mid-Atlantic Highlands; and establishing the condition of streams and estuaries through U.S. EPA's EMAP.

One important aspect of these large-scale approaches is the development of indicators that can identify areas that have the most severe problems and greatest need of management attention, action, and potentially restoration. Integration of these types of data will be challenging and will require multivariate, integrative approaches of multitaxa biological communities over large-scale landscapes and regions.

Promising new techniques to achieve integration of community measurements over multiple spatial, temporal, and biological scales are evolving such as development of multimetric indices, and statistical techniques. These techniques will require coupling with population and ecosystem-based models for aid in the interpretation of stressor risk and alternative management actions. Several large-scale programs have been initiated such as the EaGLe program reported here. For example, Danz et al. (in press) developed a stratified experimental design for the development of environmental indicators in the Great Lakes coastal region. This design was based on the compilation of more than 200 data layers on stress information for 762 coastal units and the identification of gradients of stress for several coastal ecosystem types. Six different taxa (amphibians, birds, diatoms, fish, macroinvertebrates, vegetation) were randomly sampled across these stress gradients to detect differential responses.

This work is still in progress but identifies the tremendous potential of spatially explicit public databases in the future development of environmental indicators. With the exponential increase in technological and computational capabilities including molecular techniques, remote sensing technology, modelling and statistical sophistication, data management and storage, and internet communication, analysis of biological communities as indicator signals of environmental condition and change will rapidly advance.

Coastal ecosystems have experienced tremendous stress from natural and anthropogenic influences for hundreds of years, and stress levels are projected to increase in the future. A new generation of ecologic indicators is needed to measure the condition, diagnose stressors, communicate condition to the public, assess potential future status, and evaluate management actions in our coastal regions.

Among the many challenges, the development of these indicators will require improvements in our scientific understanding of stress-response relationships, relative contributions of multiple stressors, how stressors operate over different organisational and spatio-temporal scales, and how reference conditions are determined. Fortunately, there is an explosion of ideas and technology that can aid in the multidisciplinary advancement of indicators in our coastal waters.

3

Water Rights: The Water Policy Implications

Like many rivers throughout the western United States, the Columbia has been dramatically altered by human activities. A series of major dams and diversions have radically changed the big river and its tributaries, such as the Snake, the Yakima, and the Deschutes, turning them into an economic engine for the Pacific Northwest. So thoroughly has it been exploited for hydropower, navigation, and irrigation that the Columbia has been described as a river that has "died and been reborn as money." As the Columbia River system grew more industrialised, however, its legendary salmon populations declined sharply.

The Columbia's salmon runs may once have numbered 16 million fish, but by the 1990s they had fallen to something like one million, and most of those fish were artificially produced in hatcheries. As more Pacific salmon populations were added to the national list of threatened and endangered species, recognition grew in the Pacific Northwest that the Columbia Basin ecosystem may have been pushed too far, and that changes would be needed to restore the salmon runs.

The Columbia, however, still faces new demands for water from farmers, cities, and others who continue to regard the River as a viable source. These new demands are sizable pending applications for new permits in Washington alone total up to 1.3 million acre-feet of water. These new demands could be viewed with some validity as either an incremental increase in use that is small in the context of the Columbia's annual flow, or as a new depletion that would further reduce river levels in the summertime when salmon are already stressed by low flows and high water temperatures. Irrigators and other would-be water users have argued strongly for the former view, while environmental groups and other salmon advocates have forcefully advocated the latter position.

Caught in the middle is the Washington Department of Ecology (Ecology), the agency responsible for decisions regarding new permits to withdraw water from the Columbia for use in Washington. Seeking a definitive scientific answer to the dispute over the potential impacts of new water withdrawals from the Columbia, Ecology asked the National Academy of Sciences (NAS) to study the issue. NAS, a nonprofit group of research scholars that is often called "The Supreme Court of Science," has issued a variety of influential reports on water management and the needs of native fish species, including a much-publicised 2002 draft report that questioned the scientific basis of Klamath Basin water management. Ecology's request was unusual, however, in that it came from a state agency, whereas NAS normally advises the federal government.

NAS released its report, Managing the Columbia River." Instream Flows, Water Withdrawals, and Salmon Survival (Report), on March 31, 2004. The Report finds that additional water withdrawals from the Columbia would, indeed, increase risks to salmon during critical periods of relatively low flows and high water temperatures.

These risks are exacerbated because the future of the Columbia River system contains so many big uncertainties uncertainties regarding salmon survival, water supplies (both year to year and long-term), and water demands throughout the Columbia Basin. With its finding that new water withdrawals would increase salmon risks, the Report essentially says there is no such thing as a free drink, even from the mighty Columbia.

The Report does not stop there, however. In response to Ecology's request for comment on various scenarios for Columbia River water management, NAS made a number of findings and recommendations regarding potential new water withdrawals from the Columbia. Briefly stated, the Report urges a new approach to water permitting decisions: review of new permit applications by a multijurisdiction Columbia Basin water forum, water markets and conservation as the first option to supply water for new uses, tougher requirements to obtain new permits, and greater flexibility in water allocation and management.

These recommendations may seem modest in an era when salmon recovery is a major priority of the Pacific Northwest, but they are contrary to long-standing traditions of water allocation and management in the western states. Moreover, the NAS recommendations carry major water policy implications for the Columbia Basin and beyond, because the issues and challenges facing Ecology on the Columbia are not so different from those facing water managers throughout the West.

This Chapter identifies and briefly discusses the NAS Report's water policy implications. The point is not to analyse the Report and determine if its findings and recommendations are "correct," nor to provide a detailed evaluation of the Columbia River permitting controversy and the policy response to the Report. Instead, this Chapter focuses on the NAS recommendations regarding water allocation and management, explains how they differ from the traditional approach of western states, and discusses how water management might change if the states were to adopt these recommendations. In other words, the Chapter assumes the NAS recommendations constitute "good science," and assesses how the policy toward new water demands might change if states in the West were to see the NAS recommendations as relevant and implement them.

Water Development and Management on the Columbia

The Washington controversy is fundamentally a dispute over the best use of the waters of the Columbia River. These waters have sustained salmon for millions of years and the human inhabitants of the Columbia Basin for hundreds of generations. In the 200 years since the Lewis and Clark expedition, however, the River and its tributaries have been dramatically altered, bringing many benefits for humans but also some serious costs including the decimation of the once-mighty salmon runs. Water withdrawals, primarily for irrigation, are one cause of the River's altered state; while there is sharp disagreement over the impact of existing water withdrawals on Columbia River flows, irrigation is increasingly viewed as a significant part of the problem. While water laws and institutions in the Pacific Northwest now take some account of the needs of salmon, water managers also face growing pressure to authorise new withdrawals that would further deplete flows in the Columbia. This section briefly addresses these background matters, providing some context for a discussion of NAS's Columbia River study.

The Bygone Columbia

In terms of flow, the Columbia is far and away the biggest river in the western United States, with an average annual discharge of 281,000 cubic feet per second (cfs). By comparison, the second-biggest river on the West Coast, the Sacramento, averages 23,000 cfs. The West's most storied river, the Colorado, carries less than haft the water of the Columbia's major tributary, the Snake. The Columbia drains an area of 258,000 square miles, including the vast majority of Idaho, Oregon, and Washington, plus sizable chunks of British Columbia, Montana, and Wyoming. Before it was harnessed by a number of mainstem and tributary dams, the Columbia, with highly variable flows and giant rapids, was a wild river in every sense. The untamed Columbia may have been inhospitable to large-scale hydropower, irrigation, and navigation, but it provided an enormous natural bounty of salmon. The Columbia's annual salmon runs may once have been as great as 16 million fish, a number that is roughly triple the current human population of the U.S. portion of the Columbia Basin.

These mighty salmon runs were the foundation of the Columbia Basin's first human society, that of the native tribes. Human residence in the Columbia Basin apparently dates back at least 12,000 years, and the tribes sustained themselves largely by catching salmon at Celilo Falls and other sites on the Columbia and the Snake. As stated by the Supreme Court a century ago, for these tribes, "the right to resort to the fishing places was not much less necessary than the atmosphere they breathed." Nineteenth century settlers also harvested these runs extensively, and after canning techniques were refined, salmon fishing and canning became one of the region's first major industries.

The Columbia Today, Dams, Water Use, and Salmon

Development of the Basin's water resources, and the resulting alteration of its rivers and decline of its aquatic ecosystems, occurred in three major stages. First, irrigation diversions on Columbia and Snake River tributaries such as the Boise, Yakima, and John Day Rivers began in the mid-to late-1800s.

Such diversions were often relatively small in scale but were cumulatively significant, causing many of these rivers to be severely depleted by the early part of the 20th century. Enactment of the 1902 Reclamation Act brought the second stage of development: the construction of major federal dams on Columbia and Snake tributaries to store water for irrigation. Within ten years, the United States Bureau of Reclamation (then called the Reclamation Service) was building major irrigation projects on the Boise, Snake, Umatilla, and Yakima Rivers, and many others would follow. When the Bureau of Reclamation completed a dam on Oregon's Owyhee River in 1932, it was the highest dam in the world at the time.

The third stage was the building of major dams on the Columbia and lower Snake River mainstems, largely for hydropower purposes. Construction of the Bonneville and Grand Coulee Dams began in 1933, and within 40 years there were six major federal dams on the Columbia and four on the lower Snake, along with five major nonfederal hydropower dams on the Columbia, an additional five major dams upstream in Canada, and three more nonfederal hydropower dams on the Snake at Hells Canyon.

This massive development "flattened" the Columbia River hydrograph, sharply reducing the Columbia's naturally high summer flows and increasing the low winter flows, resulting in flow patterns very different than those in which the Basin's salmon populations evolved.

Today, irrigation is easily the biggest offstream use of water in the Columbia Basin, accounting for at least 80 percent of water withdrawals for human uses. The number-two use, public water supply, is small by comparison but growing along with the region's population. Hydropower generation is the other major water use in the Columbia Basin states, accounting for more than half of the region's electrical supply and nearly two thirds of its generation capacity as of 2004.

The Pacific Northwest received many benefits from development of the Columbia River and its tributaries, but it paid with its salmon runs. Not only did dams and diversions drastically alter flows, they also blocked salmon from reaching vast areas of their historic habitat. When combined with other factors such as overfishing, logging, and other land-use practices, the effects of water development on salmon have been devastating. One leading commentator, Columbia River water expert John Volkman, has summarised the decline as follows:

The runs were still relatively large in the 1950s, but with the closing of the floodgates at the last Snake River dam, accelerated timber harvest and all the other changes, Snake River salmon populations went into serious decline. The declines became "synchronous" and widespread in the late 1960s. From historic peaks ranging from ten to sixteen million adult fish, the Columbia runs declined to something like a million. As bleak as this number is, it understates the decline of the wild salmon stocks that scientists see as the "seed corn" for the salmon run, only a handful of populations, the Hanford Reach fall chinook, the Wenatchee River sockeye and several summer steelhead stocks in Oregon's John Day River, for example, are still considered healthy.

Many populations of salmon and other anadromous fish have been listed as threatened or endangered under the Endangered Species Act (ESA). Efforts to save and restore Columbia Basin salmon populations have included some measures aimed at ensuring adequate flows in the Snake and Columbia Rivers. For example, when the Northwest Power Planning Council a regional entity charged with seeking appropriate balance between hydropower generation and ecosystem protection in the management of federal hydropower dams in the Basin developed its original plan for the Basin's fish and wildlife in the 1980s, the key element was a "water budget" to provide mainstem flows for salmon. More recently, the Bureau of Reclamation has essentially rented water from Idaho irrigators in an effort to provide an additional 427,000 acre-feet of Snake River water for salmon. Perhaps most significantly, federal agencies operating under the ESA have established flow targets on the Columbia and Snake Rivers for salmon. But Columbia River flows commonly fall below these targets, especially in the summer, largely because of water withdrawals for irrigation and other purposes. A federal study in the mid-1990s found that in the driest summers, average Columbia River flows at McNary Dam fall 90,000 cfs short of the target levels for salmon, primarily because of upstream water withdrawals for irrigation.

Water Allocation by States Under Prior Appropriation

In the Columbia Basin, as in the rest of the United States, water allocation is chiefly a state matter, and water use is regulated primarily under state law. States have occupied the field of managing water resources largely because the federal government has often deferred to the states in this area, allowing each state to establish its own rules for how its waters will be allocated and used. Federal agencies also play major roles in the management of Columbia Basin waters, especially where dam operations are concerned.

For example, the United States Army Corps of Engineers operates major dams on the Columbia, the Snake, and several of their tributaries, largely for hydropower and flood control; the Bureau of Reclamation operates dozens of dams throughout the Basin, primarily for irrigation; and the Federal Energy Regulatory Commission licenses more than a dozen nonfederal hydropower dams in the Basin, including major facilities on the Columbia and Snake mainstems. Thus, Congress has left the states in charge of allocating water among users and regulating their use, even though the federal government has been largely responsible for developing and managing the Basin's water resources, and even though these resources are shared by two nations and several U.S. states.

Washington, like Idaho, Montana, and Oregon, allocates water under a statutory system based on the prior appropriation doctrine. For purposes of this Chapter, several classic features of prior appropriation are worth noting. First, water is a public resource subject to the control of each state. Second, states recognise private rights to use this public resource if the water is diverted from its natural course and applied to a "beneficial use" such as irrigation, industry, or domestic water supply. Third, such "water rights" generally last forever. Fourth, each water right's priority is based on its seniority, so that in times of shortage, older ("senior") water rights get all the water they are entitled to before newer ("junior") rights get any. Fifth, each water right has a specific purpose of use, place of use, and point of diversion; for example, a particular water right may authorise irrigation of a specifically described 200 acres in Benton County, Washington, with water to be taken from a specifically described point on the Columbia. Sixth, a water right may be changed (or "transferred") temporarily or permanently to a new place of use, type of use, or point of diversion, but only if the state approves the change based on a finding that it would not harm other water users. Thus, an irrigator may sell or lease her water rights to another irrigator, or to a city or some other water user, but such changes may not adversely affect other users on the same system, nor can they improve a water right's priority date or increase the maximum quantity of water to which the user is entitled. The Columbia Basin states now require that any person seeking to make a new use of water (rather than just continue an ongoing use) first obtain a permit from the relevant state agency. While each state has its own requirements, in general a state agency will not issue a new water use permit if it finds that the available water supply has been fully allocated to existing users or if the permit might somehow violate the public interest. A permit provides no guarantee of a reliable water supply, however, because of the prior appropriation seniority system a permittee cannot use water if all the available supply is required to satisfy older, higher-priority rights.

Traditionally, prior appropriation did not protect water flowing in its natural course ("instream flows") primarily because the doctrine required a diversion of water from its natural course as an indispensable element of a water right, but also because of doubts about whether instream uses were truly "beneficial." Led by Oregon and Washington, the western states began easing these restrictions in the latter half of the 20th century, and today all the Columbia Basin states recognise water rights for instream flows. Instream flow protections are established differently from traditional water rights under prior appropriation because they generally are created only in response to a request from a state agency, primarily for the purpose of protecting sufficient flows to provide adequate habitat for fish populations. Once created, however, instream flow protections are much like regular water rights in that they carry a specific location and quantity of water, along with a fixed priority date that determines their place in the priority system.

Washington established instream flow protections for the Columbia River mainstem in 1980. Users with water rights senior to 1980 could continue to use water if the Columbia fell below these specified levels. Ecology continued to issue new permits, however, and these post-1980 users could have their water use cut off if the River fell below the minimum flows; these junior rights came to be known as "interruptible." In the critical drought year of 2001, however, Ecology allowed irrigators holding these interruptible rights to continue their use despite low flow conditions on the Columbia.

Through the 1980s and into the 1990s, the states continued to issue permits for water withdrawals from the Columbia and Snake as the Basin's salmon populations continued to decline. In response to ESA listings of Columbia Basin salmon populations, Idaho, Oregon, and Washington all imposed moratoria on the issuance of further water withdrawal permits from the Columbia and Snake Rivers. Although the Washington and Oregon moratoria did not absolutely prohibit all new diversions, they did represent an effort to prevent additional water withdrawals from causing further risks to salmon.

Pressure for Expanded Uses (New Rights and Uninterruptible Status)

Washington's moratorium was not popular with agricultural interests and others especially in the eastern half of the state who believed that the economic importance of new Columbia River water withdrawals outweighed any risks to salmon caused by reduced flows.

In 1997, the Washington legislature passed a bill to lift the moratorium and new Governor Gary Locke signed it, although he promised that Ecology would not "process any pending or future applications for new appropriations until instream flows are established and a determination is made that water is available for withdrawal or make decisions for new appropriations that would jeopardize regional salmon recovery efforts." Thus, lifting the moratorium did not immediately result in the issuance of new permits, but it raised the pressure on Ecology to authorise new withdrawals.

Some existing Columbia River water users were also asking Ecology to improve the status of their water rights. Specifically, those users whose water rights were junior to the 1980 instream flows on the Columbia and therefore "interruptible" during low-flow years were seeking to have their rights converted to "uninterruptible" status.

Ecology had issued interruptible water rights exceeding 172,000 acre-feet, and these users were essentially seeking to have their water rights elevated in priority over the 1980 instream flows. Serious questions persisted, however, about the potential for new or expanded water withdrawals to harm salmon populations by depleting summer flows in the Columbia River mainstem.

The National Marine Fisheries Service, responding to a proposal for a large new irrigation withdrawal from the Columbia River in Oregon, found that the new withdrawal could jeopardize salmon by further reducing mainstem flows that already fall below target levels in most years. In Washington, two environmental organisations petitioned Ecology in 1999 for a new moratorium on Columbia River water withdrawals until it established instream flow needs for salmon.

Under pressure from all sides regarding Columbia River water, Ecology launched its "Columbia River Initiative," with the stated purpose of allowing access to new water withdrawals while providing support for salmon recovery. Ecology acknowledged that decisions on Columbia River water management had become increasingly controversial, and that "lack of scientific consensus [had] resulted in gridlock."

In the hopes of resolving the gridlock, or at least providing a scientific foundation for a defensible water management regime, Ecology asked the National Academy of Sciences for a study of the relevant issues on the Columbia.

The National Academy of Sciences Study

NAS and the Committee

NAS, which conducted the Columbia River study, is the most prestigious national entity for scientific research. As NAS stated in one of its reports, "NAS and its associated organisations constitute what has been called 'The Supreme Court of Science' and the most important independent expert advisory mechanism for science and technology in the United States and perhaps in the world." The National Research Council (NRC) is the working arm of NAS, carrying out most of the studies done in its name. NRC fulfills this mission primarily by "calling upon a wide cross section of the nation's leading scientists, engineers, and other professionals, who serve on its committees without pay."

The NAS study process has multiple safeguards to provide studies that are balanced, objective, and well-grounded in science. Once a topic is approved for study, NAS seeks to ensure that each committee has the appropriate balance and expertise for its assigned task. All reports are subject to peer review for scientific validity and to further review to ensure that the committee's conclusions and recommendations are supported by the body of its report. NAS drew its committee for the Columbia River study from a variety of disciplines and backgrounds. The Chair, Ernest T. Smerdon, was an expert on water resources and agricultural development who recently retired as Vice Provost and Dean of the College of Engineering and Mines at the University of Arizona.

The other 12 committee members included a professor of agricultural and resource economics at Oregon State University; a consulting fisheries biologist from Eagle, Idaho; the director of the International School for Water Resources at Colorado State University; a consulting fisheries scientist from Redmond, Washington; a professor of social ecology with water resources expertise from the University of California-Irvine; a professor of ecology from South Dakota State University; an emeritus professor of zoology and limnology from the University of Wisconsin-Madison; a water quality expert retired from the United States Geological Survey; an engineering professor from the University of Colorado; a watershed management coordinator from the Michigan Department of Natural Resources; an emeritus professor of hydrologic science from the University of California-Davis; and an attorney with expertise in water rights and resources. The committee staff was led by Study Director Jeffrey W. Jacobs from NRC's Water Science and Technology Board. The NAS committee held a series of public meetings, three in Washington State and the last in Washington, D.C., where it heard from Ecology staff and a variety of "others with specific interests or expertise." A variety of experts from government, major universities, and the private sector reviewed the draft report. The "prepublication" final copy of the report was released to the public on March 31, 2004.

The Committee's Charge: Salmon Science and Water Management Scenarios

Ecology's requests to NAS were both general and specific. The general request was for a review of the science underlying decisions regarding new water withdrawals from the Columbia River and the potential effects of such withdrawals on salmon. The Report describes the general portion of the committee's charge as follows:

A key issue in this study was the pending applications for additional water rights permits from the mainstem Columbia River in the State of Washington, applications which have been on hold for some time. Our committee's charge was to consider the implications for [sic] potential additional withdrawals for Columbia River salmon and to comment on the body of scientific knowledge related to this issue and its implications.

The committee was not charged to review all ecological issues (of which there are many) across the basin which affect salmon but rather to conduct a more focused investigation regarding conditions in a stretch of the mainstem Columbia River in the State of Washington. Nor was the committee charged with recommending policy decisions but rather was requested to review the scientific information available by decision makers and to comment on it.

More specifically, Ecology presented five management scenarios involving proposals for water withdrawals from the Columbia River mainstem and asked NAS to respond to these scenarios. The Report notes that these five scenarios "contained many assumptions and actions related to water withdrawal quantities, management actions, and water use fees.. Several possible management actions did not contain enough specificity to enable detailed evaluation." In general, Ecology's management scenarios (except for the highly uncertain fifth scenario) would involve some combination of the following: i) issuing new permits for water withdrawal from the Columbia up to some defined quantity, ii) converting existing "interruptible" water rights to "uninterruptible" status so that use of these converted rights during low-flow periods would not be curtailed to protect flows for salmon, iii) requiring "state of the art" water use efficiency practices for new or converted water rights, iv) imposing annual water use fees applicable to new or converted water rights, v) establishing new mechanisms for permanent or temporary transfers of existing water rights into new uses, and vi) various other management measures. Ecology's five scenarios may be summarised as follows:

Scenario 1: Ecology would:

- issue new permits from the Columbia up to 1 million acre-feet (AF) per year;
- convert existing interruptible water rights to uninterruptible status;

- require "state of the art water use efficiency practices" by users of all new and converted water rights;
- reassess the management program for Columbia River water withdrawals periodically, with formal reviews at ten years and twenty years, to reflect the latest scientific information;
- establish a functioning water market or water bank for Columbia River water rights;
- assume that Oregon may withdraw up to an additional 600,000 AF from the Columbia mainstem for its future needs; and
- protect an additional 427,000 AF of Snake River instream flows through the Washington portions of the Snake and Columbia Rivers.

Scenario 2: Same as Scenario 1, except that Ecology would

- impose an annual fee of $10 per AF on all new and converted water rights;
- use proceeds of these fees to acquire water rights to improve streamflows, restore salmon habitat in the Columbia and its tributaries, and possibly develop new water storage projects;
- issue new permits from the Columbia up to 700,000 AF per year; and
- issue additional new permits up to 300,000 AF per year only upon a showing that "conservation investments were in place for a majority" of existing Columbia River water users.

Scenario 3: Same as Scenario 2, except that Ecology would:

- charge $20 per AF annually (instead of $10) for new and converted water rights; and
- provide financial support for new water conservation measures and potential water supply improvement measures, such as new storage projects.

Scenario 4: Ecology would:

- require "direct mitigation in the mainstem" for any new water use on the Columbia, i.e., require that new withdrawals be fully offset by acquiring existing water rights, implementing new water conservation projects, or using new storage;
- impose a fee of $30 per AF annually on converted water rights; and
- use the proceeds of the $30 annual fee to acquire water rights for instream flows or fund other habitat improvements in the Columbia mainstem and tributaries.

Scenario 5: Ecology would consult with state, tribal, and federal fisheries managers on each proposal for an individual new water right. "Under this scenario whether or not mitigation is required and the type and quantity of that mitigation are decisions made on each permit on a case-by-case base [sic] as a result of the consultation."

It is worth noting a couple of issues that Ecology did not ask NAS to address. First, there was no request to determine specific minimum flows of the Columbia River for salmon survival and recovery that is, NAS was not asked to quantify how much water salmon "need" in the Columbia mainstem. Second, Ecology did not ask for an assessment of the harm to salmon (if any) caused by existing water withdrawals; instead, NAS was essentially directed to take existing withdrawals as a given in determining the impacts of potential new withdrawals.

In reviewing the science regarding the impacts of Columbia River water management on salmon populations, NAS addressed a number of subjects. NAS examined trends in water flows, temperature, and quality in the Columbia River; the relationship between these parameters and salmon survival; the existing use and allocation of water from the Columbia; the laws and institutions that govern water management in the Columbia Basin; and factors (such as ocean conditions and climate change) that create uncertainty for the future because they are hard to predict and control. NAS devoted the great majority of its Report a full 160 pages to an examination of existing scientific data and other information on these subjects. This portion of the Report is summarised below in the briefest possible terms.

In its chapter on "Hydrology and Water Management," the Report notes that human activities particularly the construction of 50 plus major dams on the Columbia River and its tributaries have dramatically altered the River's natural flow patterns. Much of this change, though certainly not all, has been caused by the operation of the Federal Columbia River Power System the system of large hydropower dams on the Columbia and Snake River mainstems.

The Columbia also has a number of major water withdrawals for offstream uses, primarily for the purpose of irrigation. These existing withdrawals have little impact on Columbia mainstem flows during the winter months, but in July and August they may divert nearly ten percent of the flow in average years and more than fifteen percent in very dry years; the report stresses the need to consider the effects of water withdrawals during these low-flow periods.

The report also notes that mainstem Columbia water temperatures have been steadily increasing over the years, creating an additional risk to salmon, and that climate change could exacerbate this problem in the future. This chapter concludes by saying, "Although precise cause-and-effect mechanisms are hard to define clearly, the changes in Columbia River hydrology identified in this chapter have greatly affected the basin's salmon populations."

The following "Environmental Influences on Salmon," acknowledges that Columbia Basin salmon are among the world's most intensively studied fish species, but that much is not yet known about them. After noting that Columbia Basin salmon populations have declined precipitously over the past several decades, the Report examines research, modelling, and alternative hypotheses regarding the declines, focusing on studies regarding the effects of Columbia River flow levels and velocities on salmon survival. In the end, NAS was unable to reach a general conclusion on the importance of these factors, but found that critically low flows and high water temperatures would present a significant problem:

Within the body of scientific literature reviewed as part of this study, the relative importance of various environmental variables on smolt survival is not clearly established. When river flows become critically low or water temperatures excessively high, however, pronounced changes in salmon migratory behaviour and lower survival rates are expected.

The Report then examines "Water Laws and Institutions," highlighting the many federal, tribal, and state governmental entities that have a hand in Columbia Basin water management. It also notes that a multistate entity, the Northwest Power and Conservation Council (formerly the Northwest Power Planning Council), has been authorised by Congress to balance power production with fish and wildlife conservation in the Columbia Basin.

It goes on to discuss tribal claims to the Columbia Basin's water and salmon resources, and identifies a few of the more important federal rights and obligations regarding the Columbia (including the protection of salmon and other endangered species). The chapter ends with a discussion of state water laws and institutions, with a primary emphasis on Washington. Idaho, Montana, Oregon, and Washington all allocate rights to use water under the prior appropriation doctrine and require anyone seeking to make a new use of water first to obtain a permit from the state. The next "Better Management of Existing Water Supplies," essentially examines the prospects for meeting water demands through reallocation of existing water supplies rather than through new permits.

The chapter begins with a discussion of studies on the economic value of water in various uses such as agriculture, municipal water supply, hydropower generation, and recreation. The Report finds substantial differences in the dollar value of water based on the category of use and explains that "[t]he significance of the differences across uses is that there is a great potential to promote economic growth and increase overall social benefits by transferring increments of water between uses (from low- to higher-value uses)." It then discusses how water markets and water banks may facilitate such transfers on either a temporary or permanent basis or notes that water banks and markets are growing in popularity across the West. The chapter then notes that water conservation measures may provide a source of "new" water supply, but they tend to be expensive often too much so for many agricultural users. The chapter concludes by urging Ecology and other water management entities to promote water marketing and conservation as viable means of providing water for economic growth without the need for additional withdrawals that could harm salmon.

Water supplies in the Basin are uncertain, both from year to year (drought) and over the long term (climate change). Water demands are also uncertain, especially because there are many entities with claims or interests in Columbia Basin water but no single entity with overall management authority.

Risks to salmon grow with Columbia Basin water demands, and they are greatest when high demands coincide with naturally low flows in the River. The take-home message is that water right permitting decisions should account for major uncertainties regarding future water supplies and demands, especially given all the other factors that pose serious risks for salmon populations. As the Report states, "given the current [legal/institutional] setting and likely future climatic and other trends, additional water withdrawals from the river during seasons characterised by low flows (particularly in drought years) will pose additional risks to salmon survival," which Ecology should consider in making decisions about proposed withdrawals during such low-flow periods.

The Report goes on to address how Ecology and other Columbia Basin water entities should approach such decisions on water permitting and management.

First, it notes that an "interjurisdictional water organisation" for the Columbia Basin involving the Canadian and U.S. governments, state and provincial governments, and tribes could reduce uncertainties and risks arising from uncoordinated decision making by each of these sovereign entities.

Second, the Report suggests dealing with multiple uncertainties through an adaptive management approach whereby today's incremental management decisions not only promote the development of new information, but also preserve maximum flexibility to make future changes in response to such information.

Third, and perhaps most interestingly, the Report states that science can inform policy decisions on matters such as water permitting, but cannot make these decisions or eliminate all uncertainty surrounding them. "Progress toward 'comprehensive' management cannot be accomplished through scientific inquiry alone, but rather requires stakeholders and management agencies to work with scientists in a collaborative learning process, such as that framed by adaptive management principles."

Consistent with Ecology's request, the Report effectively offers two sets of findings and recommendations: one primarily addressing general concepts, the other reacting more specifically to Ecology's five management scenarios as described above.

Six General Findings and Recommendations

A brief "epilogue" chapter begins by noting that various human activities have affected Columbia River salmon populations and that the long-term decline of these populations is undeniable. It then identifies "six key findings and recommendations" of the Report, which may be summarised as follows:

1. Critically low flows or high water temperatures can be expected to cause significant changes in salmon migratory behaviour and lower salmon survival rates.
2. Ecology and its Columbia Basin counterparts should continue to study and consider using water transfers and market-based programs in lieu of issuing new water right permits.
3. Converting water rights to uninterruptible status is not recommended because it would reduce management flexibility in periods of low flows and high water temperatures.
4. Thorough, inclusive management strategies require not only sound science, but also readiness on the part of decision makers to act despite uncertainty.
5. Decisions on new water permits are ultimately matters of policy; but because new withdrawals would increase risks to salmon in low flow/ high temperature periods, any new permits should provide Ecology with the authority to halt withdrawals during such periods.
6. Washington and other Columbia Basin jurisdictions should establish a joint forum to document and discuss the effects of significant proposed new water withdrawals.

Reaction to Ecology's Management Scenarios

The Report also offers a few paragraphs of "evaluation and commentary" regarding the five potential water management scenarios identified by Ecology. In these paragraphs, NAS addresses a number of significant water policy issues in language that is clear, concise, and direct. Because the next section of this Chapter focuses heavily on the water policy implications of this piece of the Report, much of it appears here verbatim: In general, the adoption of concepts related to adaptive management, such as periodic review and adjustment of the program and monitoring, and market-based conservation strategies such as conservation, the use of water markets (or "banks"), and charging for water rights, is commendable.

A pervasive aspect of the scenarios is the lack of a comprehensive basin-wide assessment of water uses and needs as a context for evaluating permit applications. Small (relative to the flows of the Columbia River) withdrawal and permitted volumes will have only small, if not miniscule, effects on the water budget of the basin as a whole. All water uses accumulate, however, both in Washington and elsewhere along the mainstem, as well as the along [sic] tributary streams.

If future demands for water increase (which seems likely given recent and projected demographic and economic trends), the accumulation of risks to salmon survival will be all the greater (given the variety of risks that affect salmon survival, assigning precise and credible levels of risk to changes in flows and temperature is extremely difficult). These effects would be magnified by reductions in low flow that could attend prospective climate warming as well as during periodic unfavourable ocean conditions. The lack of a comprehensive basin-wide management structure hampers the ability to make comprehensive judgments (both in time and over space), and it supports this report's recommendation for creating a basin-wide framework for coordinating water use data and strategies.

- Conversion of interruptible to uninterruptible water rights (Scenarios 1-4). Conversion of interruptible water rights to uninterruptible status makes adaptive responses more difficult. Interruptible water rights are interruptible so that at times of scarcity, instream flows can be protected. Making any out-of-stream right uninterruptible reduces flexibility to retain water in the river when salmon need it most during periods of high demand and low flows.
- Re-evaluation at 10 and 20 years (Scenarios 1-3). The idea of re-evaluating the scenarios periodically is excellent. For this re-evaluation to be meaningful, however, the program needs to be designed so that any aspect of it could be undone (reversed) if the evaluation calls for such a reversal. No evidence is provided of any such reversibility. Instead, the result will be decreasing reversibility by allowing for some interruptible water rights to become uninterruptible. In some cases, more frequent re-evaluations might be necessary.
- Charge for water rights (Scenarios 2-4) Charges for water rights in this scenario appear to be arbitrarily selected and out of proportion to the probable costs of mitigation and the value of water to the users. For example, the scenarios specify charges of $10 to $30 per acre-foot per year to be used (among other things) to acquire mitigation water in low-flow years. This scenario thus proposes increasing the priority of a water permit for $10 to $30 per acre-foot per year and using the money to buy water for what could be several times that amount.

- Water markets (Scenarios 1-4 and perhaps 5) Water markets, water banks, and other such market-based mechanisms offer potential improvements over existing systems of water allocation. However, restricting markets only to the Columbia River's mainstem, and only to Washington, is narrowly construed. The Department of Ecology already allows for 600,000 acre-feet per year to be used by Oregon in its assumptions, but no allowances are made for uses by Idaho, Montana, or British Columbia, or by tribes.
- Structural storage measures (Scenarios 2-4 and perhaps 5). A lack of specificity in this scenario inhibits the ability to comment extensively upon it. It implies that tributaries are to be used for additional storage (which may have negative consequences for salmon), but the habitat and condition of tributaries are of critical importance for Columbia River water quality and for survival of salmon that use the river. Tributaries should thus be considered for protection and mitigation as well.

Thus, the Report has a great deal to say about water management. It addresses key scientific issues, certainly, but also touches upon law, institutions, and decision-making processes. The Report is also filled with significant water policy implications, not just for Washington, but for the entire Columbia Basin and beyond. The following section identifies and discusses some of these implications.

Water Policy Implications of the NAS Study

The heart of the NAS Report is a scientific finding that may be summarised as follows: Water withdrawals that reduce Columbia River July-August flows will increase risks to salmon, but the magnitude of the risk is hard to estimate because of uncertainty about future water supplies and demands in the Columbia Basin and about other factors that pose risks to salmon. This finding seems rather cautious and unremarkable, but it provides the scientific foundation for the NAS recommendations regarding Columbia River water management. Because these management recommendations flow logically from this scientific basis, they might seem unremarkable too except that they go directly against the western states' long-standing approach to water allocation under prior appropriation, as explained below.

A Joint Forum for Columbia Basin Water Allocation

NAS found that one of the major sources of uncertainty in Columbia River water management was the lack of predictability regarding future water withdrawals, due in part to the lack of basin-wide control or coordination regarding new water permits. Thus, NAS strongly recommended creating a joint forum among Columbia Basin water entities to consider the basin-wide implications of proposed significant new water withdrawals from the Columbia River system. The Report states,

This forum need not entail anything binding other than an obligation to refer the applications. At a minimum, proposed diversions would be subjected to professional and public scrutiny, magnitude of risk, possibilities of mitigation, and system-wide equities. A basin-wide forum for considering withdrawal permit applications would enhance unified water management across the Columbia River basin. The State of Washington and other basin jurisdictions should create a joint forum for documenting and discussing environmental and other consequences of proposed diversions that exceed a specified threshold.

This seemingly modest recommendation for regional review of individual permit applications is contrary to a basic assumption of water management in the West that each state will make its own decisions about allocation of "its" waters to serve its own interests. Of course, states have staunchly resisted any federal government interference with their individual autonomy over water allocation, especially in the context of federal environmental laws. The states have not been so vocal in opposing any potential "interference" by their fellow states, but neither have they commonly banded together in joint efforts to address water allocation issues on interstate rivers. To the contrary, states have routinely fought among themselves over interstate rivers as each state goes all out to protect its individual interests, as shown by the recent battle among states along the Missouri.

Even where states have entered into a compact that purportedly allocates the waters of an interstate river, these compacts have not precluded further litigation among the states over the management and use of the rivers' waters. The chief motivation for some major compacts was the promise of federal water projects that would allow for expanded storage and use of the waters of an interstate river but would be built only if the states could agree on allocation of its waters.

It seems fair to say, at least in the West, that states generally have viewed interstate compacts primarily as means of protecting their individual interests. By contrast, NAS has proposed that the Columbia Basin states band together for a fundamentally different purpose: to maintain the ecological commons. In other words, NAS seems to envision a joint Columbia River forum that would priorities protection of adequate flows for salmon habitat rather than allocation of water among the states for maximum "beneficial use." Thus, while state permitting statutes allow rejection of a new water withdrawal that would impair the public interest, the joint forum would essentially ask whether a new permit would harm the public interest of the entire Columbia Basin including the interest in sustaining salmon runs.

The Columbia Basin has an established regional entity, the Northwest Power and Conservation Council (formerly the Northwest Power Planning Council), charged with developing a Columbia River management regime that adequately protects fish and wildlife resources. The Report notes that the joint forum "could be convened within the existing Northwest Power and Conservation Council, which includes broad representation of political entities from across the basin," but stops short of recommending this arrangement.

The Council has been criticised for failing to take strong measures to save salmon, especially regarding mainstem flows in the Columbia and Snake. On the other hand, it was the Council which in 1993 called on the states to work together in protecting adequate mainstem flows for salmon, prompting Idaho, Oregon, and Washington to impose moratoria on new Columbia-Snake water withdrawals.

The Washington Legislature soon acted to repeal the permitting moratorium, however, which shows the difficulties of convincing any state to forego new water diversions for the good of a broader region even if the other states are willing to participate. Under pressure from the Endangered Species Act, some of the other western states have entered into cooperative agreements regarding water use and endangered species recovery in interstate river basins such as the Upper Colorado and the Platte.

The United States Department of the Interior, a key participant in the Upper Colorado and Platte recovery programs, has taken the position that cooperative regional approaches are the best way to address ESA issues involving river management. Secretary of the Interior Gale Norton has said, "We need more Upper Colorado Endangered Fish Recovery Programs and fewer Klamaths." It remains to be seen whether the Upper Colorado and Platte programs can successfully rebuild populations of ESA-listed species. These two collaborative programs do seem to demonstrate, however, that interstate agreements regarding water and ecosystem management can be workable from the states' standpoint, especially given the alternatives of conflict and litigation.

Conservation and Markets as the First Option for Meeting New Water Demands

The Report recognises that the Columbia Basin will continue to see sizeable new demands for water to supply cities, farms, and other purposes over the next 20 years demands that easily could exceed one million acre-feet in Washington alone and two to three million acre-feet throughout the Basin. The Report does not suggest that these new demands should not receive water, but it does encourage the states to consider ways to satisfy them without simply issuing new permits:

Increased flexibility in managing the Columbia River will require greater emphasis on nontraditional approaches to augmenting water supplies, such as water marketing and water transfers, and greater cooperation of political entities across the basin. These market-based programs may require capital investments in physical infrastructure and human resources investments in experts with skills in fields such as finances, marketing, and public administration. Programs such as water transfers, groundwater banking, and other measures to increase the efficiency of water use hold promise in helping sustain the regional economy in ways that do not require ever-increasing water withdrawals. Although water uses across the basin should not be simply channelled to the highest bidders for water, such measures hold promise for helping support both economic and environmental goals and should be carefully considered.

This recommendation goes against another well-established practice of the western states in allocating water: New water withdrawals should be permitted if there is water legally available for appropriation; and if there is none, the best option is to develop new water supplies through storage projects or pumping groundwater. While many observers believe that the era of major dam building in the West is over, the states have not wanted to concede this point. Water right transfers and other market-based tools have not been the preferred approach to meeting new water demands in the West, as shown by the rather small number of water transactions taking place throughout the region. And the states' approach to water conservation as a management tool is similar to Vice President Cheney's widely quoted view of energy conservation that is, water conservation may be a sign of personal virtue, but it is not a viable water policy. The Report suggests that Columbia Basin states can no longer afford simply to accept all existing water uses as a given or to rely primarily on new permits or new storage projects to meet new water demands. Instead, the states should take steps to encourage greater productivity from existing water supplies that is, to promote more efficient use of water that has already been allocated. For example, the states could subsidize improvements in irrigation systems to reduce evaporation or seepage losses, thus improving their physical efficiency. In addition, the states could help establish water banks or other market mechanisms to facilitate the (temporary or permanent) movement of water from lower to higher valued uses, thus increasing the economic efficiency of water use. Either way, existing water uses could effectively provide a source of supply to meet new demands without the need for additional withdrawals from the Columbia or its tributaries.

By emphasising water conservation and market-based approaches that would encourage (or perhaps even require) changes in established water uses, the Report runs contrary to another fundamental feature of water management in the West: nearly absolute protection of existing uses. In an earlier chapter, I observed that the Columbia Basin states have consistently refused to interfere with established water uses, even where the states have had to bend, change, or ignore traditional prior appropriation hales to do so. Of course, one drawback of protecting the status quo is that it forces prospective new users to look elsewhere for their water supply: "By refusing to apply laws that would reduce the quantity of water legally held by existing users, states limit the amount of water potentially available for new appropriations." To the extent that new demands must be satisfied by shifting water from existing uses, however, the opportunity cost of protecting the status quo goes up.

The Report identifies several areas of the West where market-based approaches are being used to address difficult water resource problems. It cites examples from California, Idaho, Oregon, and Washington, as well as a newly authorised interstate water bank on the lower Colorado involving Arizona, California, and Nevada. The Report certainly could have offered further examples, such as the thriving trade in shares of the Colorado-Big Thompson Project on the northern Colorado Front Range, probably the nation's best-developed water market. In addition, a wide variety of "water banks," many of which have at least some environmental purposes, have been sprouting across the West in recent years. In short, while the NAS recommendation may cut against a strong tradition in the western states, it is entirely consistent with a modern trend toward greater use of market-based mechanisms.

Tougher Requirements for Obtaining New Water Use Permits

Because new water withdrawals would further deplete mainstem flows in the critical July-August period and thus increase risks to salmon, NAS indicated that it did not favour new permits from the Columbia River. At least three of Ecology's proposed water management scenarios did contemplate new withdrawals, however, so the Report addresses permitting issues in some detail. One of the Report's more significant comments on permitting is the following:

A key problem in managing the basin's water is that water permitting decisions are currently made in a piecemeal fashion, with little to no consideration of their effects on other users or their degree of consistency with other decisions across the basin. If water resources and risks to salmon survival are to be better managed, Columbia River water permitting decisions must be made in a more holistic fashion, with consideration of how additional diversions would affect other users and sectors across the entire river basin.

While this passage deals in part with the need for regional decision making on new water withdrawals, it also makes a more fundamental point: Permitting decisions should consider the full range of effects that could result from new water withdrawals, including environmental impacts that may be felt both locally and regionally. Here again, this recommendation is inconsistent with the western states' traditional approach to water allocation. Much of western water law rests on a fundamental policy of maximum utilisation of scarce water resources for human "beneficial uses." States generally have followed this policy in making decisions on new water withdrawals, commonly issuing a requested permit unless no unappropriated water is deemed to be available for the new use. In addition, states have been reluctant to stop issuing new permits despite great uncertainty about whether the water resource can sustain the authorised uses. Washington's 1997 legislation lifting the moratorium on new Columbia River water withdrawals, enacted despite ongoing concerns and uncertainty about flow requirements for salmon, exemplifies this "damn the torpedoes" approach to permitting.

Even though the great majority of western states (including those in the Columbia Basin) have statutes requiring that the "public interest" be considered in permitting decisions, the Report's recommendations on this topic nonetheless seem to call for significant change in state practices. First, while state laws call for consideration of public interest factors, it is not at all clear that the states actually give these factors much weight in deciding on permit applications. While it is hard to determine the actual importance that state agencies give public interest factors in permitting, it seems significant that few if any reported judicial decisions involve a state agency's denial of a permit (or even issuance of a permit with stringent conditions) in the exercise of its power to protect the public interest. Second, assuming that state agencies do seriously consider the public interest in deciding on permits, existing laws probably limit each agency to weighing public interest factors relevant to its own state; Idaho's permitting statute is even narrower than that, providing for consideration only of the "local public interest." By contrast, the Report recommends that agencies take a broader view of the public interest in permitting, considering the implications of new water withdrawals for the entire Columbia Basin.

Instream flow programs, whereby states establish minimum flows in specific river reaches, represent an additional means of safeguarding fish habitat from the impacts of new water withdrawals. The Columbia Basin states establish minimum instream flows for protection of fish habitat, and in some areas these minimum flows have claimed the last available water in a river, effectively precluding any new withdrawals. Thus, state instream flow programs can provide meaningful protection against new depletions, but this protection is limited to those streams where instream flows have been legally created and to the specific quantities of water set aside. By calling for full consideration of the risks associated with new depletions, however, the Report suggests that these existing state programs may not go far enough in protecting fish habitat. Under NAS's recommended approach, state permitting agencies should recognise the impacts of new depletions even on streams that do not have a legally established instream flow. Moreover, the agencies should recognise that existing minimum flow levels may be inadequate, and new depletions may pose important ecological risks even if they do not reduce flows below these existing levels. Thus, the Report implies that states may need to go above and beyond their instream flow programs in order to protect fish habitat from impacts caused by new permits.

Finally, the Report discourages the states from relying on "mitigation" measures to address the impacts of new water withdrawals on salmon. NAS specifically criticizes the idea of allowing new water withdrawals that would further deplete Columbia River flows in exchange for an annual payment of $10 to $30 per acre-foot, saying that these payments would likely cover only a fraction of the probable cost of acquiring replacement water. More generally, the Report questions the effectiveness of mitigation measures, especially those that are not onsite and in kind, as a strategy for addressing the potential harms of new water withdrawals:

Although the idea of "mitigating" impacts is attractive, the reality of most mitigation measures is that they are not well coordinated; that is, a management agency may attempt to offset harmful impacts of water withdrawals in one part of a river system with mitigation measures (e.g., ecosystem restoration) elsewhere. The ultimate outcomes of such varying actions, however, are difficult to accurately predict, measure, and compare (if indeed they are ever measured and meaningfully compared, which they often are not), thus making it difficult to determine if "mitigation" was actually achieved.

In the end, the Report acknowledges that permitting decisions are matters of public policy, and that new permits may be issued for Columbia River water withdrawals. In making permitting decisions, however, the states should consider all the ecological risks and other impacts that may result from new water withdrawals, and ensure that mitigation measures effectively address those impacts. Such requirements would make new permits less certain and more expensive to obtain, leaving them a less attractive option for prospective water users. Of course, this result is fully consistent with the Report's suggestion that water conservation and markets, rather than new permits, should be the first option for satisfying new water demands in the Columbia Basin.

Greater Flexibility in Water Allocation and Management

The Report emphasizes that Columbia Basin water managers face enormous uncertainty regarding future water supplies and demands, while salmon populations (and those responsible for managing them) face even greater uncertainty. It recommends that water managers respond to these uncertainties by employing an adaptive management strategy that is, by taking incremental steps to generate new information and by maintaining the flexibility to make changes in response to such new information. The Report concludes, however, that state water allocation and management under prior appropriation has been too rigid, and that better flexibility in these areas is crucial:

Water withdrawal applications and permitting decisions are highly contentious in both the State of Washington and other parts of the Columbia River basin. Inflexibilities in traditional western U.S. prior appropriation doctrine have contributed to these tensions. A greater degree of flexibility in traditional water permitting and rights processes is paramount to better water management and to decreasing tensions and conflicts in the basin. This report recommends implementation of a joint basin-wide water management forum and the pursuit of nontraditional water marketing and conservation strategies. A water permitting and rights process that more explicitly recognises seasonality of flows should also be devised. Decisions regarding the granting of new water rights are issues of public policy, but additional water withdrawals during the critical high demand and low-flow periods discussed in this report will increase the risks of survival to listed salmon stocks. It will also decrease the flexibility of management institutions to allocate water between different uses in critical low-flow conditions. To increase the flexibility of water management organisations and programs and to better recognise uncertainties regarding future supplies and demands, a new permitting process should be created that allows for withdrawals to be discontinued during periods of low flow and periods of comparatively high water temperature.

The kind of flexibility the Report recommends is the exact opposite of what western water laws and water allocations seek to provide maximum certainty for water users. Several basic elements of western water rights offer a high degree of security for water users: Water rights are private property rights (even though water is a public resource by law), they last forever provided they are regularly used, and they hardly ever contain any sort of "reopener" provision allowing the states to impose significant new restrictions for the purpose of protecting public values. Although the bedrock principle of "beneficial use without waste" could be applied to cut back excessive water uses under existing rights, a state's invoking this principle would create potential uncertainty for users and that is surely a major reason why states have rarely used this authority. Water right adjudications, proceedings which attempt to determine all existing water rights in a river basin, are unbelievably long and complex proceedings, but many western states are now pursuing them in an effort to ensure greater long-term certainty for their water users. Secretary of the Interior Gale Norton has emphasised that western water law (and water users) value certainty above all else: If we yield to this temptation [to resolve water disputes based on politics or public opinion rather than established water law], we will destroy the one thing that everyone who needs water seeks the most certainty and predictability about how it will be allocated in times of shortage. Without certainty, people cannot plan for their future. Without certainty, it will be far more difficult for market-based transfers to succeed. Without certainty, we will not be able to protect our environment. It is at these times that we must renew our commitment to resolving disputes and answering tough questions within the framework of enduring principles of law.

Thus, while NAS clearly recognises that its call for flexibility would represent a significant change in the western states' traditional approach to water allocation, in reality the Report calls for a truly fundamental shift. The management scenarios proposed by Ecology would provide very little of the kind of flexibility recommended by NAS. These scenarios emphasize 1) the conversion of interruptible water fights to uninterruptible status and 2) the approval of permits for new water withdrawals, with certain provisions for water conservation and mitigation. The scenarios make no provision for limiting these allocations to a certain period of years or for revising them at a later time based on new information or changed conditions. The scenarios did propose that Ecology revisit its Columbia River water management program in ten years and twenty years; NAS endorsed this idea, but noted that such a re-evaluation would be meaningful only if any aspect of the program could be reversed based on the re-evaluation. "No evidence is provided of any such reversibility. Instead, the result will be decreasing reversibility by allowing for some interruptible water rights to become uninterruptible." Thus, NAS correctly saw a basic conflict between maintaining flexibility to make changes needed to protect salmon, and providing certainty to water users through permanent, uninterruptible water rights.

The simple premise of the NAS recommendations is that salmon should not bear all the risk associated with new water withdrawals in the Columbia Basin. Thus, the Report urges states to avoid permanent, fixed water allocations in favour of an adaptive management strategy: "Decisions and policies should promote flexibility while their outcomes are being evaluated and better understood." NAS observed that water banks, water markets, and incentives for water conservation all "tend to support greater water management flexibility and merit careful consideration." If new permits are issued, the Report recommends preserving flexibility by including specific conditions to allow withdrawals to be discontinued during critical periods. The NAS Report did not mention some other permit conditions that could improve flexibility, such as reopener clauses (allowing a change in permit terms based on some future contingency) or time limitations (requiring the applicant to apply for renewal after a certain number of years). Some people might read this NAS recommendation as giving certainty to salmon and shifting all the risk to water users, but that interpretation seems unrealistic.

In theory, adaptive management has been a key principle of Columbia Basin salmon recovery efforts since 1984. In practice, the agencies overseeing these efforts have generally been quite cautious, avoiding initiatives that would fundamentally change resource allocation and use to the detriment of key industries. If state agencies commit to the principle of adaptive management of Columbia Basin water resources a big "if"they could implement it only by taking actions opposed by water users and their allies. Water use interests remain powerfully influential in every western state, as demonstrated by the 1997 Washington statute lifting the Columbia River permitting moratorium. Adaptive management will work in this context only if politics allows it to work, and that, too, is a big "if."

With its recommendations for a joint Columbia Basin water forum, water markets and conservation as the first option to supply water for new uses, tougher requirements to obtain new permits, and greater flexibility in water allocation and management, the Report clearly has major water policy implications. It remains to be seen, however, whether the Report will actually bring about water policy changes in Washington or anywhere else. The concluding section of this Chapter addresses this question.

One of the most notable features of water policy in the West is its resistance to major changes. Because NAS recommends significant changes in the western states' usual approach to water allocation and management the kind of changes that many states have at best been slow to adopt one must ask whether the Report is likely to have any significant water policy impact.

The Report is most directly relevant to Washington, of course, the state that requested the NAS study. Because the Report was not released until March 31, 2004, it was too early as of this writing to determine how it might affect Washington's management of the Columbia. On the other hand, the reaction of key stakeholder groups has been both unmistakable and unsurprising: environmental groups love it, while agricultural interests hate it. One irrigation group told state officials that their characterisation of the Report's findings regarding the risk to salmon from new July-August water withdrawals "reflects a fanatical desire to support a 'no net loss' water policy, an extraordinarily ill-founded basis for your 'water bank' proposal."

The agricultural community's harsh criticism of the Report shows how difficult it is to resolve contentious water issues through science, or even to gain acceptance of a National Academy of Sciences study by those who oppose its conclusions.

Ecology's initial response to the Report came in the form of a discussion proposal unveiled on April 30, 2004. Ecology identified two major goals, "meet the water supply needs" and "reduce risks to fish," requiring 500,000 to 750,000 acre-feet per year in the period from April through September; one-third of this water would be allocated instream for salmon, and two-thirds offstream for human consumptive uses.

To provide this water, Ecology proposed a state-sponsored water bank, with water made available from conservation measures, storage projects (existing and new), and water right acquisitions. Funding would come from a mix of state and private sources, including fees for new water users. Six months later, Ecology produced both a draft administrative rule and a draft bill regarding Columbia River water management. These draft documents, like the initial proposal, call for Ecology to acquire water for the Columbia mainstem from a variety of sources, to allocate two thirds of this water for consumptive uses and one third for instream flows, and to use both public and private money to fund these acquisitions.

Thus, the agency proposed a program that is consistent with the NAS recommendations favouring conservation and markets over new permits, although the proposal ignores the recommendation for regional review of permits and is somewhat contrary to NAS's call for greater flexibility in water allocation and management. As of early 2005, Ecology's plans were to accept comments on the proposal through May 12 and adopt a final rule in June 2005.

Will the Report directly affect water policy in states beyond Washington? That seems unlikely, at least for now. The Report focuses specifically on the Columbia, especially the Columbia mainstem, so its findings and recommendations are most relevant for Idaho and Oregon, the two states (other than Washington) that lie predominantly within the Columbia Basin. These states have already established rules that seriously restrict the issuance of new permits in the Columbia River system, so they may believe that their policies are already consistent with the Report. Moreover, there has been no significant push to establish the kind of joint Columbia River water forum recommended in the Report.

This situation could change if one of the Columbia Basin states begins to authorise major new water withdrawals; such action would cause would-be water users elsewhere in the Basin to turn up the pressure on their own state governments to ease restrictions on new water allocations from the Columbia and its tributaries. Thus, if one state begins to issue new Columbia permits (despite the opposition of environmental groups and other salmon advocates), others may well follow suit. At that point each state could be expected to face questions on whether its water policies were truly consistent with the NAS recommendations, and the proposal for a regional water forum may seem more ripe for discussion. Thus, if the floodgates or more aptly, the headgates open on major new Columbia River water withdrawals, the Report and its recommendations may gain added significance throughout the Pacific Northwest.

Western states outside the Columbia Basin might dismiss the Report as irrelevant; every river system is unique and every species has its own biological requirements, so the science of water and salmon in the Columbia does not translate directly to other places. Moreover, most western states might disregard any study of the Columbia because they see it as a bigger river with bigger ESA issues than may be found in the more and parts of the West. While the Columbia certainly has more water than other western rivers, it shares many of their common characteristics: Its natural flow patterns have been dramatically altered by human development; its waters have been extensively allocated for offstream uses, primarily irrigation; its management is shared by numerous state, federal, and tribal entities; it provides habitat for fish and wildlife species protected by the ESA; and it faces new demands for water for municipal, agricultural, and other uses.

In these respects the Columbia is a classic western river, only bigger. And given NAS's finding that new water withdrawals would raise significant ecological concerns on the mighty Columbia, it might well identify equal or greater concerns on smaller rivers that provide habitat for rare fish or wildlife species if, of course, it was asked to study the issue. Given that many states in the West would resist implementing the kind of recommendations that NAS made for the Columbia, the states might be reluctant to make such a request.

To the extent that the NAS findings and recommendations are meaningful beyond Washington, might they indirectly affect water policy in other states? In other words, if other states regarded the Columbia River report as "good science" relevant to their own water resource issues, might they consider adopting its recommendations regarding water management in interstate river systems, water markets and conservation, requirements for new permits, and flexibility in water allocation and management? The most realistic answer seems to be no, given the western states' evident lack of interest in significant water law reform since at least the 1980s. On the other hand, there are at least three reasons to believe that some western states might be open and responsive to this type of scientific input in making decisions on water allocation and management.

First, there is now an important precedent: Washington asked NAS for a study of its water management options on the Columbia a remarkable step by a western state. Just as significantly, Ecology responded to the study with a proposal that was largely consistent with the NAS recommendations. It is certainly true that Washington is just one state, that its water policy is in some respects more progressive than that of other western states, and that its Columbia River management policy is still a work in progress which ultimately may not follow the recommendations. On the other hand, Washington is a fast-growing western state with a powerful agricultural community and a water code based on prior appropriation. If science can have a real water-policy impact in Washington, it potentially can do so elsewhere.

Second, the western states have repeatedly supported the use of science, and of adaptive management principles, in managing their natural resources. For example, the Western Governors' Association has espoused a set of principles for environmental management and decision making, one of which advocates the use of "a range of respected scientists and peer-reviewed science" to help establish basic facts and determine key areas of uncertainty.

More directly relevant are the "Park City Principles," developed under the leadership of the Western Governors' Association and the Western States Water Council to guide future water resources decision making in the West. One of these principles states that water policies must be able to adapt to changing conditions, needs, and values; accommodate complexity; and allow managers to act in the face of uncertainty.

This principle recognises the complex implications of the holistic approach to water management. Policies, and the managers who implement them, must provide flexibility to adjust as needed while ensuring predictability for those involved in and affected by water decisions. As decisions become more complicated, managers may want better and more certain information and guidelines. That will not always be possible, and the system must recognise and allow managers to act in the face of uncertainty.

While not mentioning science per se, this principle is generally consistent with the Report's endorsement of an adaptive management strategy for water resources that recognises the need to make decisions despite uncertainty and relies on science to test hypotheses and inform decision makers.

Thus, at least in concept, "good science" is a key part of the western states' vision of good resource management. Third, science is just one of many factors that are nudging the western states toward changes in their standard approach to water allocation and management.

The other factors include the increase in species listed under and protected by the ESA; the enormous population growth under way across the West; the ongoing economic transition, as agriculture and other traditional industries account for an ever-smaller slice of the region's economic pie; and the growing demand for environmental and recreational amenities, many of which rely on an adequate supply of water. None of these factors is primarily scientific in nature, but they all have exerted some pressure on the states to revise their water policies and practices, often (though not always) in ways that are consistent with the NAS recommendations. Thus, while science by itself is not powerful enough to bring about water policy reform at the state level, it may add some important influence to the nonscientific forces that create pressure for such reform in the West.

How would a state allocate water based on the NAS recommendations for good management? It would engage other states and other governmental entities, including tribes and federal agencies, in making decisions about new water withdrawals. It would look to existing water rights to provide the necessary supply for new uses through a combination of conservation projects and market-based transfers. It would consider potential impacts and risks to a range of public values, beyond the prospective economic benefits, in making decisions on new permits. And if new permits were issued, they would provide flexibility to make meaningful changes in the future as necessary to respond to changed conditions or new information.

Such proposals are certainly not new. Similar ideas have appeared in various water policy prescriptions over the past several years, such as the Park City Principles, the report of the Western Water Policy Review Advisory Commission, the report of the Longs Peak Working Group, and most recently the Interior Department's Water 2025.

In general, such recommendations for water policy reform have had only minor practical effects on water policy in the western states. With the new Report, however, the recommendations have now been grounded in extensive scientific studies and delivered by an objective and highly credible scientific source. In a policy arena where conflicting values often result in gridlock, that scientific credibility should count for something. Of course, unlike the United States Supreme Court, NAS does not get the final word in matters of law. Nonetheless, "the Supreme Court of Science" speaks with a strong and authoritative voice, and its Columbia River Report may thus have a significant influence on water policy in Washington and beyond.

The Electricity-water Nexus and the Electric Utility Sector

During the drought of 2002, lawmakers in Idaho ruled that five large coal and gas-fired power plants should be denied water rights for cooling because they would deplete much needed freshwater for drinking and irrigation. In Nevada, the 1,580 megawatt (MW) coal-fired Mohave Generation Station was forced to close in 2005 due to lack of groundwater. A few years earlier, American National Power had to withdraw its application to build a 1,100 MW natural gas plant near Hillburn, New York, because it created a controversy concerning water rights. Far from being isolated examples, water issues have complicated power plant construction or operation in Arizona, Georgia,

California, Colorado, Massachusetts, Missouri, New Mexico, North Carolina, Pennsylvania, Rhode Island, South Dakota, Tennessee, Texas, and Wisconsin.

Figure: *Important connection in the energy-water nexus—agriculture.*

The situation underscores a problem as pressing as it is invisible to many electric utilities, water planners, and even ordinary people: burgeoning water use at conventional thermoelectric power plants. Water use for electric power plants increased five-fold from forty billion gallons per day in 1950 to 195 billion gallons per day in 2000. The average power plant in the United States uses about twenty-five gallons of water for every kilowatt hour (kWh) generated. Given that electric utilities produced 4,159,514 gigawatt hours (GWh) of power in 2007, these power plants ostensibly used 104 trillion gallons of water. This amount is enough to cover the entire country in two inches of water or to almost completely fill Lake Erie.

This chapter explores the consequence of the growing water needs of the U.S. electric utility industry, and suggests that lack of water during the summer months in many regions could complicate continued reliance on thermoelectric power plants that combust coal, oil, natural gas, and biomass (or utilise nuclear fission) to generate power. The importance of exploring the electricity-water nexus, its associated challenges, and its possible remedies is threefold.

First, a slew of government agencies and industry groups, including the National Research Council, U.S. Geologic Survey, U.S. Department of Energy, U.S. Department of Interior, Electric Power Research Institute, Sandia National Laboratory, National Energy Technology Laboratory (NETL), and the National Renewable Energy Laboratory, have recently issued reports focusing on the importance of water use at conventional power plants. These reports, however, mostly argue that better technologies will need to be developed in order to address the industry's growing water needs. The National Research Council calls on the federal government to increase research and development (R&D) funding for innovative energy technologies that utilise less water, while both Sandia and NETL discuss treating and reusing brackish water, capturing water vapour from power plants, and diffusion driven desalination as important technical options. Another study argues that the President should issue an Executive Order granting the Federal Energy Regulatory Commission (FERC) authority to designate select parts of the country "National Electricity-Water Crisis Areas." In December 2008, the Supreme Court heard arguments in Entergy Corporation v. Riverkeeper for and against strengthening the Environmental Protection Agency's regulations concerning the intake of water at conventional power plants, but the Court has not yet made its decision and the case concerns only the damage of power plant intake structures to fish and aquatic biodiversity. Senators Bingaman and Murkowski even introduced legislation in early 2009 to commission a study on electricity-water problems and produce a roadmap, but their bill is uncertain to pass and again focuses on federal action. While focusing on federal research and national legislation is indeed important, equally significant is the role that electric utilities, public utility commissioners, and state regulators can undertake to avoid new thermoelectric power plant construction, invest in energy efficiency and renewable power resources, and alter electricity prices, either in conjunction with or independent of federal action.

Second, despite this collection of reports, existing electricity planners and water managers do not appear to be responding to electricity and water problems quickly or comprehensively. The Clean Air Task Force has also concluded that "water use and consumption have not been significant factors in decisions related to the permitting and siting of power plants." Peter Gleick has noted that "energy and water issues are rarely integrated into policy." Neither energy nor water planners are consequently trained to think about electricity and water in a systematic way. Electricity industry advocates continually downplay the importance of clean energy sources for minimising thermoelectric water consumption. When assessing solutions to reduce the water use from conventional generators, the Electric Power Research Institute, for example, mentions variable speed electric drives, advanced membranes, ozone disinfection, electroseparation, and freeze-thaw wastewater treatment as important water technologies, but not energy efficiency, demand side management, or renewables.

Conversely, an influential RAND report on water management focuses exclusively on tools such as "supply management" (including the location, development, and exploitation of new sources of water such as building new dams and control structures, desalination plants, arranging for inter-basin transfers of water, reclamation and reuse) and "demand management" (such as water quality matching, privatisation, and water pricing) but never once mentions energy policy tools or more efficient power plants. This chapter is an important call for more synergistic thinking that views electricity and water problems as interconnected, especially insofar that energy efficiency and renewable power plants can simultaneously reduce demand for electricity and improve the availability of water.

Third and finally, the challenges related to water scarcity and electricity are not confined to the United States. The Central Intelligence Agency believes that more than three billion people will be living in water-stressed regions around the world by 2015 (with a majority concentrated in North Africa and China). Water tables for major grain producing areas in northern China are dropping at a rate of five feet per year, and per capita water availability in India expected to drop fifty to seventy-five percent over the next decade. The American Museum of Natural History reports that about 900 million people spread across twenty-seven developing countries already lack adequate access to water. Thus, an exploration of how utilities in the United States may respond to electricity-water crises can offer policymakers insight into how the industry can address what is sure to become a pressing global dilemma.

The Electricity Water Nexus

Almost all conventional power plants, including coal, oil, natural gas, and nuclear facilities, employ one of three types of cooling cycles in their generation of electricity. Once-through cooling systems withdraw water from a source, circulate it, and return it to the surface body. Re-circulating or closed-loop systems withdraw water and then recycle it within the power system instead of discharging it. Dry cooling systems, which are not widely adopted, use air instead of water to cool power stations.

As their name implies, once-through cooling systems, or "open-loop" systems, only use water once as it passes through a condenser to absorb heat. After it passes through the plant, heated and treated water is then discharged downstream from its point of intake to a receiving body of water. Since such cooling systems release heated water back to the source, they can further contribute to evapourative loss by raising the temperature of receiving water bodies. Once-through systems are responsible for withdrawing ninety-one percent of the nation's water used for power plants, and are also utilised by more than half of the country's fleet of nuclear reactors.

Re-circulating or closed-loop systems, by recycling water, withdraw much less of it but tend to consume more. To maintain plant performance, water is frequently discharged from the system at regular intervals into a receiving body of water or collection pond, but is otherwise recycled as much as possible. Since it is being reused, the water requires more chemical treatment to eliminate naturally occurring salts and solids that accumulate as water evaporates. Closedloop systems also rely on greater amounts of water for cleaning and therefore return less water to the cooling cycle.

Figure: *An emblem of the energy-water nexus: falling.*

Dry-cooling, an approach that replaces evapourative cooling towers in closed-loop systems with cooling towers dependent entirely on air, works best in colder weather and in arid environments. Only a very small number of plants rely on dry cooling, since they lower plant efficiency and cost the most.

When taken together, the once-through, closed, and dry cooling systems in place at these power plants use a significant amount of water. (The term "use" encompasses both water consumption and water withdrawal). Using the most recently available data from the U.S. Geologic Survey, thermoelectric power plants used more than 195 billion of these gallons of water per day, or fortyseven percent of the nation's total, in 2000. According to the U.S. Geologic Survey, water withdrawals for thermoelectric generators differ greatly by state. When looked at geographically, Texas withdrew the largest amount of water; Illinois, Texas, and Tennessee accounted for twenty-two percent of all total freshwater withdrawals; and California and Florida accounted for more than forty percent of saline surface water withdrawals. This means that, on average, thermoelectric generators use more water than the entire country's agricultural and horticultural industry.

Such immense water needs produce equally immense concerns given the likelihood of future droughts and shortages, especially during the summer months. Even under normal conditions, water managers in thirty-six states anticipate shortages in the next ten years. Similarly, using a historical database of droughts going back to 1895, the U.S. Geologic Survey has predicted that almost one-fourth of the country will risk severe droughts by 2040. The most severely hit part of the country will be the West. As early as 2025, the U.S. Department of Interior cautions that "demands for water in many basins of the West exceed the available supply even in normal years." Given the intensity of the existing electricity industry's water needs, if it grows as predicted, it will directly trade off with the water available for drinking, industry, and agriculture. Utilising data from the U.S. Census Bureau, U.S. Energy Information Administration, U.S. Geologic Survey and National Oceanic and Atmospheric Administration, it appears that these tradeoffs will become most severe in twenty large metropolitan, areas. These regions of the country expect to add at least 2,700 MW of thermoelectric capacity and will experience population growth of at least 500 people per square mile. Thus, these regions will face water shortages of at least 1.52 inches in the summer by 2025.

Electricity Water Challenges in Eight Metropolitan Areas

To better illustrate many of the challenges facing electric utilities and water planners in these metropolitan areas, this section explores eight of them in detail: Atlanta, Charlotte, Chicago, Denver, Houston, Las Vegas, New York, and San Francisco. These eight regions plan to add a collective 129,828 MW of thermoelectric capacity between 2000 and 2025, power stations that would use 25.6 trillion gallons of water per year (or more than seventy billion gallons of water per day).

Figure : *The greatest global water challenge according to new research.*

The dynamics of the water needs and consequences for each region will differ greatly. In some regions, such as Colorado, Georgia, and North Carolina, greater water use from power plants in 2025 could compete with the water needed for drinking, industrial manufacturing centres, and commercial enterprises. Water withdraws and consumption for power plants in Illinois and Nevada could deplete water from Lake Michigan and Lake Mead (respectively,) violating international law. In California and New York, greater power plant additions could compromise the stability of fisheries and accelerate the extinction of endangered species.

In the Atlanta metropolitan area, Georgia Power, a subsidiary of Southern Company, intends to add 3,480 MW of thermoelectric capacity between 2000 and 2025. Georgia Power currently services 1.13 million customers in the Metro Atlanta region, yet their 16,000 MW portfolio is heavily water-intensive. About seventy-five percent of their fleet is coal powered, eighteen percent nuclear powered, six percent oil and gas powered, and one percent from hydroelectric sources. Indeed, more water will be lost as steam from Georgia Power's two nuclear plants than used by all residents of downtown Atlanta, Augusta, and Savannah combined. Within the state as a whole, thermoelectric plants use slightly more than half of all surface water, which then reduce drinking water supplies by reducing flows to Lake Lanier.

The most immediate consequence of increased thermoelectric water use in Atlanta will be tradeoffs with other major industrial and commercial water users in the region. These include Georgia-Pacific Corporation (one of the world's largest manufacturers of tissue, packaging, paper, pulp and building products), Mohawk Industries (the world's largest producer of flooring and carpets), and the city's water utility. The top commercial Atlanta customers for water included plants operated by die Coca-Cola Corporation, Pepsi Cola Corporation, Lockheed Martin Corporation, and Edwards Baking Corporation. Pepsi's Gatorade plant, for instance, uses about five million gallons of water every month.

State policymakers seem to recognise the danger of water shortages, and a fierce legal battle has erupted. Georgia is fighting to hold back more water along its river basins and reservoirs, but Florida and Alabama argue that Georgia has mismanaged water resources and that extra Georgian withdrawals would dry up river flows that support out of state power plants, farms, fisheries, and industrial users. Alabama, for example, says that restrictions on water use in Georgia would impede electricity production at their Farley Nuclear Plant, also on die Chathoochee River, threatening power outages among 800,000 residents in three states. Tri-state water negotiations have so far only precipitated into eight active lawsuits, and Georgia's state assembly passed a resolution calling on the governor to set up a commission looking into having the border redrawn through die middle of Chattanooga, Tennessee. Resolutions were later introduced in both the state House and Senate to annex part of Tennessee to increase Georgia's access to water. The mayor of Chattanooga, who would lose half his city if Georgia's border was redrawn, sent a consignment of water bottles to Georgia lawmakers. He publicly announced it was better to "offer a cool, wet kiss of friendship radier than face a hot, angry legislator gone mad with thirst."

In Charlotte, Duke Energy Corporation reported plans to add 17,950 MW of diermoelectric capacity between 2000 and 2025. Duke Energy, the primary electric utility, relies on more than 28,000 MW of electricity capacity to meet the needs of 3.9 million citizens of Ohio, Kentucky, Indiana, North Carolina, and South Carolina (die Carolinas account for 2.3 million of its customers). Of its total capacity, thirty-nine percent is coal-fired, thirty-seven percent is nuclear, thirteen percent is hydroelectric, and eleven percent oil and natural gas meaning that their entire portfolio is water-intensive.. Duke already announced in March 2008 that it needed to import 520 MW of power outside of the region to ensure continuation of service during a prolonged drought.

One town in North Carolina was so dry mat water had to be imported by fire truck. In 2008, die Summer nuclear plant (near Columbia) was at such a "critical point" that operators openly discussed having to shut it down for lack of water. The Harris nuclear reactor near Raleigh obtains water from Harris Lake, which was a scant 3.5 feet above the limit that the plant could operate. Most of Duke's power plants draw their cooling water from the Santee River Watershed, which includes the Catawba- Wateree River, a water source that has earned the title "America's Most Endangered River for 2008.

This "Most Endangered River" basin, however, is about the only place to situate new thermoelectric power plants, and the associated water use with these capacity additions could exacerbate drought (at best) and risk interstate litigation and agricultural collapse (at worst). Low water levels along the river leave bottom dwelling organisms such as clams and mussels stranded, and tend to induce algal blooms that contaminate water supplies. Low water levels also degrade drinking water infrastructure, since they diminish revenues for water utilities (meaning they have less earnings available for maintenance,) and dry and crack soil leading to pipeline malfunctions and spills.

In Chicago, electric utility planners intend to add 40,386 MW between 2000 and 2025. Chicago is formally served by Commercial Edison, a subsidiary of Exelon Corporation. Exelon, one of the largest electric utility providers in the nation, serves 5.2 customers, but under a $4.8 billion deal sold many of its power plants to Midwest Generation (which generates electricity but only sells it on the wholesale market). Commercial Edison, now the nation's largest supplier of nuclear power (and the world's third largest supplier), operates one of the most water-intensive fleets in the electricity industry. The total amount of thermoelectric power generation in the area is about 12,649 MW, drawing mostly from a mix of local rivers, Lake Michigan, and Powerton Lake.

The most immediate impact to future thermoelectric withdrawals from Lake Michigan could be a violation of domestic and international law. The Supreme Court decided in 1996 that the State of Illinois had to limit its diversion and use of water from Lake Michigan. Under the ruling, Illinois must legally reduce its water usage over the next fourteen years - not increase it for power plants. Furthermore, since the Great Lakes contain twenty percent of the world's freshwater supply and ninety-five percent of the freshwater for the United States, they are strenuously protected under international law.

The 1909 International Boundary Waters Treaty, signed between Canada and the United States, specifically governs water resource use on Lake Michigan. The most recent agreements, signed by all Great Lakes State Governors and Canada's Provincial Premiers in December 2005, recognise that "the Waters of the Basin are a shared public treasure and the States and Provinces as stewards have a shared duty to protect, conserve and manage these renewable but finite Waters." Thus, deploying more thermoelectric power plants could defy domestic court rulings and international treaties.

In the Denver metropolitan area, Xcel Energy may build 4,503 MW of thermoelectric capacity between 2000 and 2025. 'Xcel operates three power generating facilities in Denver - the Cherokee, Zuni, and Arapahoe Plants - all which use water from the South Platte River. Located south of downtown, the Arapahoe Station is a 156 MW coal-fired plant that uses about 586 million gallons of water annually from the South Platte River.

The Zuni Station, closer to downtown, is a 107 MW natural gas plant that uses 98 million gallons of water annually from the South Platte River. The Cherokee Station, north of downtown, is a 717 MW coal-fired plant that uses 2.4 billion gallons of water from the South Platte River annually (along with about 5,200 acre-feet of recycled water from Denver Water's Reuse Water Plant).

Because the South Platte River is the primary water source for the region, increased power plant withdrawals and consumption could seriously deplete the water needed for households and businesses (about forty-seven percent of Denver's water is used by single family homes). Water use from any additional power plants would have to be balanced against the needs of the many downstream users along the South Platte. A severe shortage could risk economic and environmental impoverishment, since the municipal drinking water system in Eastern Colorado and farming and ranching activities in Nebraska rely on the river.

In Houston, electricity demand could grow to need 26,989 MW of new thermoelectric capacity by 2025. The situation with retail electricity access in the Houston area is more complicated than many other cities (although the construction and dispatch of generation remain roughly the same). In June 1999, Texas Senate Bill SB 7 made retail electric competition state law and phased-in "retail choice" of electricity on January 1, 2002." The legislation forced incumbent Texan utilities such as TXU and Reliant Energy to restructure then assets into separate retail marketing, generation, transmission, and distribution companies.

The bill attempted to lower prices and limited the market power of existing suppliers by unbundling transmission, distribution, and generation; it also prohibited any single utility from owning more than fifteen percent of generation within a service region.

Thus, the electricity market serving Houston is complex. More than 200 different electricity plans are available to Houston residents and service is offered from sixty-eight competitive retailers, including eight large utilities Champion Energy, CPL Retail Energy, Direct Energy, Gexa Energy, Green Mountain Energy, Reliant Energy, Spark Energy, and TXU Energy that generate and distribute power from five distinct zones all over the state. Taken together, these power providers operate about 95,400 MW of capacity, forty-eight percent fuelled by natural gas, thirty-nine percent by coal, eleven percent from nuclear, and three percent from "other" sources.

The largest of these is the "wires only" utility Centrepoint Energy, which maintains the T&D infrastructure serving 5,000 square miles around the city (although it does not generate a single kwh of power). Centre Point serves two million customers in Houston and operated 3,600 miles of transmission lines, 226 substations, twelve service centres, and 43,000 miles of distribution lines to deliver 75.9 billion kwh in 2006.

Increasing the number of power plants in the Houston vicinity, however, could reduce the water needed for drinking and agriculture. In the past, Houston depended primarily on groundwater to meet eighty percent of its supply, but rapid depletion has lowered that amount to only sixty-seven percent today, forcing the city to take more water from the Trinity, San Leon, and San Jacinto Rivers (along with the reservoirs they support). With Houston water planners predicting rising demands for drinking water, there may not be enough water for new power plants.

Surface water upstream from Houston is also needed to irrigate agriculture. During the last serious water shortage caused by a prolonged drought in 1996, the agricultural sector was the first to suffer as water was diverted to supply power plants and drinking water systems.

Las Vegas, Nevada

Population is growing so quickly in Nevada that more than 20,000 MW of thermoelectric capacity may be built in Las Vegas between 2000 and 2025. Because of the rapid growth in population, Nevada Power Company, the primary electric utility, is also one of the fastest growing utilities in the country. Nevada Power's service area covers 807,000 customers spread across 4,500 square miles in southern Nevada, including the cities of Las Vegas, North Las Vegas, and Henderson. It operates 35,990 miles of T&D lines and has invested $5.4 billion in about 5,623 MW of mostly natural gas and coal-fired power plants, which it uses to meet 54.3 percent of the region's power needs. The shortfall is made up from purchased power imported from Arizona, Utah, and other parts of the state. The water needs for this portfolio of power technologies is significant. Nevada Power's fleet of plants currently use water from a combination of local reservoirs, underground aquifers and wells, and the Colorado River, with a significant amount coming from the beleaguered Lake Mead. According to water use profiles reported by the Nevada Division of Environmental Protection, the 465 MW El Dorado gas power plant uses 176,100 gallons of water per day from Lake Mead.

The almost completed Copper Mountain Power Plant, a 660 MW natural gas plant soon to be operating in Clark County, will use 75,000 gallons of potable water per day from Lake Mead. The Reid Gardner Station, a four-unit 650 MW coal fired power plant in the Moapa Valley, uses groundwater that would otherwise replenish Lake Mead. The power plant consumes about 286,000 gallons of water per day from the Muddy River (which flows into Lake Mead), but has also contaminated local aquifers and wells. The Reid Gardner station's cooling ponds and waste landfills have leaked sulfates and toxic pollutants into the watershed. Groundwater underlying the power plant has become so contaminated that it exceeds federal safety standards for no less than twelve criteria pollutants including arsenic, chromium, lead, and selenium.

If more thermoelectric power plants are built as planned, the most immediate impact would be greater water shortages for the Las Vegas Valley Water District and further reductions in Lake Mead, which is already experiencing and "extreme hydrological drought."" Researchers from the Federal Bureau of Land Reclamation estimate that the Lake Mead water system is losing 326 billion gallons of water per year, or enough to supply eight million people, a loss so significant it can be seen by satellite images from space.

If new power plants further tap Lake Mead, the result could be a full blown agricultural disaster. Lake Mead directly irrigates about a million acres of farmland in southern California's Imperial Valley, "and another half million acres in northern Mexico as part of [an] international water treaty."

New York, New York

Planned thermoelectric capacity additions for New York could surpass 11,600 MW by 2025. The major electric utility directly serving New York City is Consolidated Edison, which provides electric service to most of the City (excluding a small part of Queens) and Westchester County. The utility operates about 1,739 MW of its own capacity (mostly oil and gas plants) to supply 10.5 percent of its power, but imports the rest from competitive suppliers through an integrated bulk power system known as the New York Control Area. About half electricity generated within the city comes from power plants in Queens County, and a majority of this power comes from the 2,450 MW Ravenswood natural gas power plant on the East River. Consolidated Edison takes its generated and purchased power and delivers it to about three million customers through 94,000 miles of underground electric cables - the largest underground system in the world and enough cable to wrap around the Earth more than three times.

The water needs for the portfolio of power generators that must supply power within Consolidated Edison and the New York Control Area are substantial, but they do not tradeoff directly with the City's water supply. The City's water comes from nineteen reservoirs, three controlled lakes, and about 300 miles of aqueducts spanning the Catskill Mountains to Westchester County. Instead, new power plant additions would likely consume and withdraw water from the Hudson River, where seven facilities constituting 6,691 MW of capacity - Bethlehem Energy Centre (previously the Albany Steam Station), Danskammer Generating Station, Rosetón Power Plant, Indian Point Energy Centre, Lovett Power Plant, Bowline Power Plant, and the LRT Power Plant on 59th Street - already use 6.1 billion gallons of water directly from the Hudson for coolant every day.

New power plants could consequently have a devastating impact on local fisheries and ecosystems through the discharge of heated effluent, entrainment, and impingement. A series of extensive fishery surveys along the Hudson, including the Long River Survey (assessing egg and larval densities), Fall Shoals Survey (assessing densities of juvenile fish populations), and the Beach Seine Survey (assessing the abundance of fish communities) determined that thermoelectric power plants were devastating freshwater fisheries in the early 1970s.

The federal government passed extensive regulations to limit the damage, but utility restructuring this past decade has renewed concern that electric utilities, more focused on competition, will be less focused on environmental compliance. Indeed, one 2006 study has already warned that additional thermoelectric power additions along the Hudson could increase mortality of Striped Bass (already weakened from current power plants), bay anchovy (already suffering fifty percent year class reductions due to power plants), and Atlantic Tomcod (at stress due to any further increases in water temperature). A similar 2007 study warned that the withdrawal of cooling water for new thermoelectric plants would have: "profound impacts on aquatic environments" [including] "reductions of phytoplankton, zooplankton, fish, and shellfish, including the loss of threatened and endangered species; damage to aquatic organisms, including important elements of the food chain; reduction of populations and their compensatory reserves; losses of commercial and recreational fisheries; and stresses to overall communities and ecosystems."

San Francisco, California

In San Francisco, electricity demand could grow to justify the construction of 4,759 MW of thermoelectric capacity between 2000 and 2025. Both competitive power producers and electric utilities operate in the San Francisco Bay Area. The Mirant Corporation, an independent power producer, operates the 674 MW Contra Costa natural gas plant, the 1,311 MW Pittsburg natural gas plant (both situated near the confluence of the Sacramento and San Joaquin Rivers), and the 362 MW Potrero natural gas plant (which uses cooling water directly from the San Francisco Bay). Pacific Gas and Electric Company (PG&E) also operates the 530 MW natural gas-fired Gateway Generating Station near Antioch. The plant receives its water from the Delta at the confluence of the Sacramento and San Joaquin Rivers. PG&E's 2,200 MW Diablo Canyon Nuclear Power Plant, located in Avila Beach, uses water from the Pacific Ocean.

Given the region's unique hydrology, the immediate effect of additional water use from new thermoelectric power plants could be drought, shortages of drinking water, and species extinction. Most of power plants in the Bay Area draw water from the Sacramento-San Joaquin Delta, a massive, natural, inverted river delta, where numerous waterways converge downstream, eventually flowing into the Suisun Bay and the Upper San Francisco Bay (where they proceed to flow into the Pacific Ocean under the Golden Gate Bridge). The same water system replenishing the rest of California, therefore, is the one that power plants predominately use.

The water system is instrumental in distributing water to twenty-five million Californians extending beyond San Francisco through the Central Coast to Los Angeles and San Diego. About 2.5 million acres of productive farmland rely on the Sacramento-San Joaquin Delta for irrigation, and the California Draught Preparedness Organisation has already warned that water supplies are running dangerously low. Further contributing to a shortage of water, a 2007 District Court ruling protecting the endangered Delta Smelt effectively cut California's drinking and water irrigation supplies by thirty-five percent. Almost one month after the court decision, the Coalition for a

Sustainable Delta filed a lawsuit against the Mirant Corporation, alleging that their Contra Costa and Pittsburg natural gas power plants harmed the Delta Smelt by using more than a billion gallons of water a day to cool steam turbines. In April 2008, further restrictions on water withdrawals were implemented out of concerns that rapid declines in Delta water were harming the Sacramento River Chinook Salmon. New conventional power plant additions, it appears, could risk possible species extinction in addition to drought and water shortages.

Electric Utilities and Regulators

What can regulators in these cities and states, along with the eight utilities most at risk and the dozens of others possibly like them, do in the face of these electricity-water challenges? While many disputes over water in the West and Southeast are interstate, local regulators and electric utilities are still exceptionally positioned to respond to such risks within their jurisdiction. City and state regulators are often much more attuned to the needs and dynamics of their constituents than federal policymakers, and can respond more quickly and efficiently to local problems. Electric utilities serve as direct intermediaries between electricity consumers and power providers and generators; have established supply,metreing and billing relationships with their customers; are more flexible than government agencies and can quickly tailor programs; and possess the most knowledge about electricity production and use in their area.

Figure : *Regulators and buswork inside an electric utility substation.*

This section argues that while a cornucopia of different technologies and mechanisms are available to regulators and electricity utilities, a combination of six could be very effective at avoiding future water shortages: increasing research and development funding for alternative power plant cooling cycles, placing a moratorium on thermoelectric power generation, strongly promoting energy efficiency and demand-side management, rapidly deploying wind turbines and solar photovoltaic panels, changing electricity prices and giving electricity customers more feedback and information, and asking the federal government (or forming interstate organisations) to designate select regions of the country "Electricity-Water Crisis Areas."

Increase Funding for R&D on Alternative Cooling Cycles

The U.S. Department of Energy, through separate programs at the National Energy Technology Laboratory and Sandia National Laboratory, has begun to look into making conventional power plants more water-efficient, and a number of emerging technologies can greatly reduce water use. Researchers working at these programs, for example, have investigated treating and reusing "impaired," "nonpotable," "produced," "brackish," "reclaimed," or "gray" water to cool power plants.

The most common applications include using secondary treated municipal waste water, passively treated coal mine drainage, and ash pond effluent. Fifty-seven power plants, mostly in the arid Western part of the United States, already rely on cooling cycles that utilise reclaimed water, and while abundant sources of reclaimed water are available in Alaska, California, Kansas, Louisiana, Oklahoma, Texas, and Wyoming (accounting for 90.1 percent of produced water), it would be a prohibitively costly for forty-three states. Of the produced water in these seven states, only thirty-seven percent is of the quality needed to run in power plants.

Another option being explored is enabling power plants to produce some of their own water, either through capturing water vapour from flu gas or using the thermal discharges from power plants to desalinate water. Water is naturally present in all deposits of coal, constituting as much as sixty percent of its weight. The coal combustion process thus releases water vapour which can be recovered from flu gas using liquid desiccant-based absorption systems or modified electrostatic precipitators. Engineers at NETL expect that such capture technologies could reduce five percent of evapourative water loss at power plants. Diffusion driven desalination, a process that uses the excess waste heat from power plants to produce distilled water, can also minimise the water needs of power plants situated in coastal areas.

Thermoelectric power plants, in other words, need not always consume and withdraw water as they do today, and opportunities exist to substantially improve the efficiency of thermoelectric cooling cycles. While none of the technologies discussed here are yet available at a commercial scale, doubling or tripling the funding for the research programs at Sandia and NETL could provide the technological breakthroughs needed to greatly reduce water use at power plants in the future.

Place a Moratorium on New Thermoelectric Power Generation

Perhaps the simplest response electric utilities can take is to stop building new thermoelectric generation in areas where water shortages are expected to occur, or water prices anticipated to rise rapidly. As the two next sections suggest, existing thermoelectric and hydroelectric capacity would be sufficient to meet existing power needs but energy efficiency, wind, and solar could offset any expected increases in electricity demand.

Using these three tools to offset thermoelectric capacity additions is important because drought and flood are a normal, recurring part of the North American hydrologie cycle. Even though meteorological droughts, identified by a lack of measured precipitation, are difficult to predict and can last months to decades, every part of the country has experienced severe or extreme drought conditions at least once since 1896 - with about half of the country suffering drought conditions ten to fifteen percent of the time. Given the likelihood of future water shortages, utilities could justify transitioning away from water intensive thermoelectric generation.

The addition of new conventional power plants therefore has two inherent water-related risks that suggest electric utilities should no longer construct them: they are unable to withdrawal water needed for normal operation in times of scarcity, and can cause and worsen existing water shortages when their fuel cycles consume water. Thermoelectric power plants running on coal, natural gas, oil, and uranium require immense amounts of water to cool the combustion process. They withdraw trillions of gallons of water from our rivers and streams. They consume billions of gallons of water from local aquifers and lakes. A moratorium would therefore offer a very effective tool for preserving water resources.

One possible objection to a moratorium would be that future increases in electricity demand can only be reliably met by fossil-fuelled and nuclear baseload power plants. While this concern is a legitimate one, the next two sections show that the promotion of energy efficiency, demand side management, and improved feedback to electricity customers could offset the need to build any new thermoelectric capacity.

Furthermore, attaching wind turbines to pumped hydro and compressed air energy storage systems can improve their capacity factor to above seventy percent, making them "functionally equivalent to a conventional baseload plant." Paul Denholm found that generating base-load power from a hybrid wind/compressed air energy storage system would add only about 0.70/kWh to its cost of producing electricity, and that converting natural gas plants to biomass generators to backup wind farms would add only about 0.2 0/kWh, two relatively inexpensive options to create completely renewable base-load units. Contrary to some proclamations stating otherwise, "a baseload wind system," notes Denholm, "can produce a stable, reliable output that can replace a conventional fossil or nuclear baseload plant."

Similarly, Vasilis Fthenakis and his colleagues from Brookhaven National Laboratory, the Renewable Energy Research Institute, and the Institute for Analysis of Solar Energy recently concluded that with new advances in compressed air energy storage make it "clearly feasible to replace the present fossil fuel energy infrastructure in the [United States] with renewables." Researchers at the Georgia Institute of Technology and the Virginia Polytechnic Institute & State University even found that when coupled with a rigorous energy efficiency and demand management program, solar panels could completely displace the electricity currently coming from the two GW Indian Point nuclear facilities in New York at a cost less than building a new nuclear power plant.

Promote Energy Efficiency and Demand-Side Management

To offset the risks of placing a moratorium on future thermoelectric generators, electric utilities should rigorously implement energy efficiency and demand-side management programs (DSM). Curiously, such action would not only help address impending electricity related water shortages, but would also improve energy security, lower electricity and water prices, and enhance reliability. For the historical record suggests that energy efficiency, DSM, and load management practices represent the most feasible way of responding to increases in electricity demand. Increasing energy efficiency, one study concluded, "is generally the largest, least expensive, most benign, most quickly déployable, least visible, least understood, and most neglected way to provide energy services." Or, as Jon Wellinghoff, the Commissioner of the FERC, put it, "the potential benefits from the incorporation of demand response into wholesale markets indicate that a considerable margin of gain is possible from accelerating such activity."

The Department of Energy (DOE) recently calculated the benefits of DSM and found that it lowers wholesale electricity prices as costly power plants are displaced and total demand on the system decreases. Generating peak electricity is extremely expensive, often exceeding $5,000 to $10,000 per installed kW (meaning a 100 MW plant can cost $750 million to build and require seventyfive million dollars per year to operate), implying that DSM should be profitable for all utilities. Eric Hirst estimated the costs and benefits of DSM programs for three types of utilities: a "surplus" utility, an "average" utility, and a "deficit" utility.

Surplus utilities are those with excess capacity and few planned retirements, as well as slow projected growth in fossil fuel prices and incomes". Deficit utilities have little excess capacity, many planned retirements, and rapid growth in prices and incomes. Average utilities fall in the middle. Hirst found in each case that DSM programs raise electricity prices but reduce electricity costs, and the overall percentage reduction in cost far exceeds the increase in price. He found 2-l cost benefits for the surplus utility (i.e., every dollar spent on DSM yielded two dollars of savings), 5-l cost benefits for the average utility, and 8-l cost benefits for the deficit utility. In other words, energy efficiency may cause electricity prices (per kWh) to go up slightly, but because people and companies use fewer kWhs, their bills actually go down. For some utilities, cost savings can be as high as eight dollars for every one dollar invested in DSM.

For these reasons, a few states have experimented with large-scale DSM programs. In the state of New York, cost-effective DSM and load management policies have saved more than 1,000 gigawatt hours (GWh) of electricity and displaced 880 MW of peak demand in the past five years. Montana, Idaho, and Oregon conserved energy totalling more than 50,000 GWh per year in the mid-1990s with a retail value of two billion dollars to consumers. California's DSM programs operating between 1990 and 1992 delivered 112 percent of planned energy savings, meaning they cost-effectively saved more than electricity than intended. The Massachusetts Department of Energy Resources, discovered their load management program lowered participants' electricity costs by twenty million dollars in 1999, with benefits to Massachusetts customers exceeding six million dollars in just thirteen hours on one high-cost day.

Notwithstanding these impressive gains, there is much more potential in energy efficiency and DSM than some ever imagined. The National Association of Regulatory Utility Commissioners (NARUC) found cost-effective energy efficiency potential in all regions of the country, with the most untapped potential in the Northeast and South, where electricity costs are highest (meaning energy efficiency efforts are more economical than areas where energy is cheaper). Another study projected that cost-effective energy efficiency programs could reduce consumption by around one trillion kWh by 2020, offsetting almost all projected growth in electricity use - and the needed capacity additions to achieve it. The Alliance to Save Energy found that aggressive investments in energy efficiency could free up enough electricity to eliminate the need to construct more than 1,300 power plants in the next twenty years. One study projected that a national DSM program aimed at reducing peak demand by just five percent would yield three billion dollars in net generation, transmission, and distribution savings per year and displace some 625 infrequently used peaking plants and associated delivery infrastructure. Other broader studies confirm DSM's cost-effectiveness.

The International Energy Agency reviewed forty large-scale commercial DSM programs found that they saved electricity at an average cost of 2.1 to 3.0 eVkWh. Similarly, the Institute of Electrical and Electronics Engineers found an average cost of 2.6 0/kWh for demand-side management, load management, and energy efficiency programs in Vermont166. When thinking about these numbers, readers are asked to consider that one kWh saved is more valuable than one kWh generated. One thousand MW of energy efficiency and 1,000 MW of large-scale generation are not equivalent. Richard Cowart found that every dollar invested in energy efficiency:

- Mitigated against uncertainty and lowered load, wear, and maintenance needs on the entire fossil fuel chain, even in hours when reliability problems were not anticipated by system managers;
- Depressed the costs of locally used fuels such as oil, coal, and natural gas;
- Reduced demand across peak hours, the most expensive times to produce power;
- Lessened costly pollutants and emissions from generators;
- Improved the reliability of existing generators;
- Moderated transmission congestion problems;
- Operated automatically through customers coincident with the use of underlying equipment or load, meaning they are always "on" without delay or the needed intervention by system operators to schedule or purchase the resource.

Discounting all of these benefits and looking purely at reductions in peak demand, the New York ISO determines a reserve criterion of eighteen percent during times of peak demand to ensure overall system reliability. Accordingly, each 1.0 MWh of peak demand that customers avoid through energy efficiency means that utilities can subtract 1.18 MWh of total capacity needed. Quite literally, every single kWh of peak electricity avoided through energy efficiency equates to 1.18 kWh of avoided supply. While this calculation would not necessarily hold true for the displacement of base-load generators, since most peaking power plants rely on natural gas the water savings could still be substantial. Natural gas units are the third most waterintensive of all power plants, using about fourteen gallons of water per kWh (second to coal at thirty-six gallons per kWh and nuclear power at forty-three gallons per kWh).

Furthermore, a majority of energy efficiency savings can be accomplished with small and targeted programs. In the ISO-New England service area, for example, about nine percent of the system's total generating capacity is tapped one percent of the time. The price of power during this most expensive one percent accounts for sixteen percent of the total annual dollars spent on the spot market. These numbers reveal that a very small fraction of supply accounts for a large fraction of total cost. Nationwide, Cowart calculated that just 0.4 percent of industrial customers account for thirty percent of total demand. Consequently, relatively small DSM programs directed at a miniscule proportion of the nation's electricity customers can produce mammoth benefits in terms of total demand reductions.

Finally, a significant amount of energy efficiency and DSM potential exists in the eight metropolitan areas identified above. Georgia Power could cut projected load by thirty-three percent for residential and commercial customers (and twenty-six percent for industrial customers) by 2018. The same study found an achievable potential for energy efficiency programs - that is, energy that could be saved if existing programs were funded to their full potential - of 2,509 GWh per year. Another assessment conducted by the Governor's office estimated that twenty-three to twenty-four percent of cumulative load growth could be displaced by cost-effective energy efficiency programs (amounting to the equivalent of about 1,771 MW of displaced capacity).

In North Carolina, Duke Energy has acknowledged that their "save-a-watt" program could displace 1,318 MW of peak capacity by 2012 if properly funded. Similarly, Lawrence Berkeley National Laboratory researchers estimated that Chicago had at least 3,456 MW of untapped DSM potential. Environment Illinois, a nonprofit group, concluded that home weatherisation programs, enforcement of energy star appliance standards, and light bulb replacements could cost-effectively save the state more than 2,100 GWh, displace the need for fourteen billion cubic feet of natural gas, and cut home energy use by twenty percent. If coupled with more aggressive retrofitting schemes and other investments, household energy consumption would be reduced by forty percent (and with technological improvements, ninety percent). In Colorado, Xcel Energy admitted that it could ramp up its DSM program to provide an additional 200 GWh (equivalent to eighty MW of peak potential) if properly incentivised.

An independent report from KEMA found that the state of Colorado at large had at least 2,031 MW of technical energy efficiency potential. Texas has 40,000 MW of energy efficiency potential characterised by demand response, solar PV, and improved building codes, appliance standards, and public energy efficiency initiatives. These alternative sources could meet 107 percent of the projected growth in summer peak demand by 2013 and displace thirty-three percent of peak load by 2023. A DSM program in Texas would not only displace conventional generation, but would also result in utility bill savings of seventy-three billion dollars at an avoided cost of 4.50/kWh, much cheaper than the state-average retail electric price of 9.IeTkWh. And, in Nevada, fully funding just nine energy efficiency programs could displace the need to construct 5,571 MW of capacity in the Sierra Pacific and Nevada Power service areas by 2020. In New York, the National Research Council, along with a study led by researchers at the Georgia Institute of Technology, found a significant amount of programmable energy efficiency potential within New York City. That is, existing programs would only need to be properly funded to achieve roughly 4,461 GWh of potential by 2012.

Both studies show that residential energy efficiency improvements in lighting, cooling, refrigeration, electronics, space heating, and hot water heating, along with commercial improvements in lighting, refrigeration, cooling, ventilation, office equipment, water heating, space heating, and building controls, could cost-effectively displace the need to build 550 MW of peak capacity by 2015.

The studies estimated that an additional 300 MW of peak capacity could be displaced by aggressive demand response programs. The Natural Resources Defence Council and Pace University Law School Energy Centre (now the Pace Energy & Climate Centre) estimated even greater potential in the New York City area, and concluded that savings of up to 3,032 MW peak demand could be achieved by aggressive energy efficiency programs within two years. Finally, the San Francisco Bay Area Economic Forum calculated that scaling up load management programs in the region could displace 2,500 additional MW of thermoelectric capacity.

Deploy Wind Farms and Solar Photovoltaic (PV) Panels

In situations where energy efficiency and DSM programs are unable to completely offset the need to construct new thermoelectric power plants, utilities could rely on wind turbines and solar panels to produce electricity.

These two technologies use almost no water to generate electricity, and need only a very small amount for cleaning and maintenance. Even more remarkably, looking at the marginal levelised cost of building power plants in 2009 - that is, the cost of constructing, operating, maintaining, and fuelling a new facility - offshore and onshore wind turbines produce electricity for between 2.6 and 5.6 0/kWh, making them two of the six cheapest sources of power. Solar PV is the most expensive at 39 0/kWh, but not far behind expensive peaking plants that cost between 32.5 and 35.6 0/kWh to operate. Wind, in other words, is already cheap, and solar (which is getting cheaper) is nearing parity with natural gas peaking facilities.

Nationally, commercially available wind and solar photovoltaic power generators could provide almost 3,000 GW of capacity - roughly three times the country's existing capacity - by 2010 if only they were built. (Building these wind and solar power facilities would also have the additional benefit of providing high-paying manufacturing jobs in those regions of the country hardest hit by the financial crisis). Again, while the following estimates and projections should be taken in context, the eight metropolitan regions discussed in Part II have substantial amounts of wind and solar potential.

In Georgia, solar PV systems could provide at least 104 MW of cost effective peaking power by 2010. In Atlanta, 13,223 MW of peak capacity could be displaced by placing solar arrays on just commercial and residential buildings. Moreover, certified wind maps project more than 10,000 MW of offshore wind power could be imported to Atlanta. In North Carolina, an independent consulting team found 9,600 MW of onshore wind potential, with a "practical" amount of 1,500 MW, that could be installed in Duke Energy's service territory to produce 3,900 GWh of electricity pear year. Indeed, researchers at Stanford University and the University of Delaware projected that offshore wind energy near the Mid-Atlantic coast could supply 330,000 MW of power - enough to meet the energy needs of nine states from Massachusetts to North Carolina (including D.C.) with electricity left over to support a fifty percent increase in demand.

In Chicago, the DOE confirmed that Illinois and the region surrounding the city had at least 9,000 MW of potential onshore wind power at only five of the state's prime sites. These areas represent about 1.2 percent of state land and each square kilometre of windy land could support about five MW of installed wind capacity. Offshore wind potential in Lake Michigan is even greater. A report partially funded by the Wisconsin Focus on Energy Program, after sampling hundreds of wind sites and assessing lake depth, concluded that the mid-lake plateau would be "excellent" for offshore wind farms with steady and strong winds, and that the Lake has energy potential "in excess of 10,000 MW." The estimate was confirmed by an independent and interdisciplinary analysis of potential offshore sites which found that 9,700 MW could be developed.

Colorado's combination of high mountains and flat plains creates a bounty of wind resources, with strong and consistent winds coursing through the western part of the state and Denver. The Governor's Energy Office has noted that a significant amount of this wind - about 93,000 MW - could be converted into electricity. The DOE found more potential, and noted that the state has about 170,000 MW of potential wind capacity (with the capability to produce 288 billion kWh of electricity per year), even when making exclusions for land use and environmental permits. The Hewlett Foundation also projected that eighty-three million MWh per year of solar PV resources could be deployed within the state, or about twice as much electricity as consumed state-wide. Three of the five largest wind farms in the nation are located in Texas: the state holds the record for the world's largest wind farm (the 736 MW Horse Hollow Wind Energy Centre), and it is the site for the nation's second-largest wind farm (the 505 MW Sweetwater Wind Project). The potential, nonetheless, has hardly been exhausted, and the Texas Public Utility Commission estimates that as much as 25,000 MW of wind power could flow to urban centres by 2012. The Union of Concerned Scientists projects that die state could add another 17,800 MW of wind by 2025.

In Nevada, after assessing land use patterns and favourable wind sites, researchers discovered a large concentration of achievable wind energy near Las Vegas.

The University of Nevada calculated that state had the potential for more than fifty million MWh of wind power, or the equivalent of 3,240 MW of installed capacity in the three counties surrounding Las Vegas. Solar PV resources for Southern Nevada are excellent: about 7,000 to 7,500 watts hours per square metre, making Las Vegas "one of the best sources for this type of generation in the world." Researchers at the Hewlett Foundation and the Energy Foundation projected that the tri-county area surrounding Las Vegas could provide 11,700 MW of power from solar PV systems. New York City has

an equally impressive amount of building integrated solar PV potential. Researchers at the Centre for Sustainable Energy at Bronx Community College projected that solar PV systems mounted on rooftops and building façades could supply eighteen percent of the city's electricity by 2022. Each square foot of New York City receives the equivalent of 160 kWh of sunlight per year, enough commercial and residential roof space to host between 8,500 MW and 15,700 MW of PV installations. Researchers at the University of Albany estimated that 2,431 MW of solar PV potential exists for the state of New York on large parking lots alone.

Lastly, the San Francisco has estimated that it has at least thirty MW of small-scale wind generation in the Bay Area. The nonpartisan California Energy Commission (CEC) concluded that the Bay Area could rapidly deploy 4,000 MW of wind potential in the form of large onshore wind farms in only a few years. Stanford University researchers projected that between twenty-six percent and 12 percent of California's entire electricity needs could be provided with offshore wind. A single 2,385 MW wind farm sited on the Northern California coast, for instance, could produce about 9.7 TWh of electricity, enough to displace 5.6 percent of the state's fossil fuelled generation (the authors comment that the 9.7 TWh estimate is on the low end of the potential for the Bay Area, with a high end of twenty-seven TWh).

Researchers at the Golden Gate University Law School estimated annual PV output for San Francisco at 550 to 1300 GWh per year. At the high end, this is comparable to current residential (and industrial) electricity use in San Francisco. At the low end, the researchers estimated that PV output could displace 600 MW of peaking power plants. The researchers noted that over the thirty year period that they analysed, the standard deviation of annual average solar radiation was less than 2.5 percent of its average value, suggesting that solar resources would be very reliable. They also noted that these PV systems could be integrated into buildings and rooftops, meaning they would have no new requirements for land. The CEC analysed the state's solar PV resources and found that since they could be deployed virtually anywhere in the state, their technical potential exceeded seventeen million MW of capacity. If applied only to existing residential and commercial rooftops, the CEC concluded solar PV potential exceeded 74,000 MW of capacity.

Change Electricity Prices and Improve Information

A moratorium on thermoelectric generation, energy efficiency and DSM programs, and wind and solar PV deployment should be supplemented with utility efforts to alter electricity prices and provide feedback and information to electricity customers. Three changes would be most significant: more accurate electricity pricing, altered electricity billing practices, and a utility-wide information program to educate consumers.

Consumers are generally unaware of daily, weekly, and seasonal changes in electricity price, and instead see only a monthly electricity bill. They thus consume electricity indiscriminately. Utility programs that reflect time-of-use through "real-time," "interval metreing," "time-of-use", or "seasonal" rates could show customers how electricity production and consumption varies according to the time of day, week, and month. Most electricity bills combine charges into a lump sum, making it difficult for consumers to tell how much of the bill results from the individual use of appliances or technologies, how much the bill could be decreased by using more efficient models, or how much electricity use can be shifted to off-peak times.

The Energy Policy Act of 2005 (EPAct) implicitly recognised this flaw in electricity pricing, and encouraged utilities to provide time-based rate schedules reflecting variations during the day to all individual customers requesting it. However, the EPAct also said, ambiguously, that each state regulatory authority can decide whether to implement that provision. Correspondingly, Lawrence Berkeley National Laboratory estimates that only about 100 utilities (less than two percent) offered some sort of time-of-use rate for electricity customers in 2007.

Yet a preponderance of evidence suggests that pricing electricity more accurately will greatly improve the efficiency of the electricity industry, provide customers with proper price signals, and reduce wasteful energy use. One study provided residents with daily electricity prices for a month and found a 10.5 percent reduction in electricity use. Another analysis of residential electricity use from 1973 to 1980 found that "feedback" in the form of information detailing daily and weekly electricity prices reduced consumption between six and twenty percent. When Princeton University researchers gave residents of Twin Rivers, New Jersey, information about their level of electricity and natural gas use on a daily basis, consumption dropped ten to fifteen percent. Another study involved eight experiments tracking electricity use at 602 households over the course of many years. In some experiments, feedback was given three to four times a week, and in one experiment it was given continuously and informed households of the cost of their consumption every half hour. The researchers found that frequent, credible feedback about electricity prices resulted in 10-13 percent less electricity use than control groups.

In two of the metropolitan areas discussed above, New York and Atlanta, electric utilities have already started altering electricity prices to change behaviour. New York has experimented with "alternative rate designs" by offering time-of-use rates, day-ahead real time pricing, critical peak pricing, and pricing at real-time market rates. Researchers at LBNL surveyed 149 commercial and industrial customers in the Niagara Mohawk Power Corporation service area where the utility offered time-of-use tariffs for large customers with peak demand needs. They found that more than thirty percent of industrial customers responded by foregoing discretionary electricity usage and fifteen percent shifted usage from peak-periods to off-peak periods; forty-five percent of respondents installed demand reduction enabling technologies on site; and peak load for the utility was reduced by fifteen percent. Further south, Georgia Power introduced time-of-use metres for large industrial customers and, from 1992 to 2002, enrolled 1,650 customers to reduce peak demand by seventeen percent.

Because most people remain uninformed about the electricity-water nexus, a second form of feedback could be useful: making water usage associated with electricity generation "visible" by including it in people's electricity bills. California, for example, was the first state in the country to enact an Advance Recovery Fee on sales of some electronics. Whenever customers purchase new cell phones and televisions, a visible fee between six and ten dollars appears separately on receipts, showing consumers how much it will cost to collect and recycle some of the toxic elements inside their products. The same technique could be used in electricity billing, where consumers could be shown a separate line estimating the number of gallons of water used (and/or its associated cost) to produce the electricity they used within their home that day, week, or month. The last type of information that utilities can offer consumers is education. The DOE noted in 2008 that eighty-eight percent of Americans would fail even a "basic" electricity literacy test. Another recent report in Kentucky found that forty-one percent of respondents identified coal, steel, and oil as "renewable resources." Because they remain uninformed and misinformed about electricity, many consumers will not understand the scope and scale of the electricity industry's water needs. Utilities could fund and promote electricity information and education campaigns to teach the public about electricity and water use. These could include grade-school classes on energy and the environment; public demonstrations and tours of renewable power facilities; free energy audits and training sessions for industrial, commercial, and residential electricity customers; and utility sponsored document "clearing houses" that offer websites, free books and articles to help consumers gather and process information in order to make more informed choices about their electricity use.

Designate National "Electricity- Water Crisis Areas"

Finally, each of the above efforts could be complemented by state or federal designation of national "Electricity-Water Crisis Areas." Cities and states expecting the most severe water shortages and thermoelectric capacity additions could form an interstate organisation to monitor and manage electricity-water resources, or could approach the federal government (most likely the FERC) to establish a National Electricity Water Policy Program Office through either Congressional legislation or an Executive Order.

This Program Office could advise and assist local and state water managers and energy policymakers in their efforts to implement the efforts described above. The Program Office could chair a cabinet-level National Electricity Water Policy Interagency Task Force consisting of water and energy policy experts from the national laboratories, the DOE, the Department of Agriculture, the Department of Interior, the EPA, the Office of the Federal Environmental Executive, the Council on Environmental Quality, the Federal Energy Management Program, and the U.S. Geological Survey. This Task Force would have a few benefits over state-level action, since it could have the authority to coordinate and harmonize federal laws to stimulate the expedited implementation, permitting and siting of clean power facilities in all federal lands within or adjacent to National Electricity-Water Crisis Areas.

The pervasive water challenges confronting the electric utility industry in the United States suggest that local regulators, public utility commissioners, water managers, and electric utility operators start immediately responding to them. The water needs for expected future thermoelectric power plants - if they continue to use twenty-five gallons of water per kWh as they do today - in just eight metropolitan areas are so immense they could deplete the water available from Lake Lanier in Georgia and exacerbate interstate litigation between Tennessee, Alabama, and Florida by 2025. Biodiversity could perish along the Catawba-Wateree River Basin in North Carolina. Chicago could find itself embroiled in domestic and international legal disputes over the consumption and withdrawal of water from Lake Michigan. Households and businesses could run out of water from the South Platte River in Colorado. Rivers could stop recharging the groundwater needed for drinking and irrigation in Texas. Lake Mead and the Colorado River could continue to suffer drought in Nevada, inducing an agricultural crisis in California and Mexico. Fisheries along the Hudson River in New York could collapse. The Delta Smelt could become extinct in the San Joaquin River Basin in California. These impending but avertable risks serve an important reminder that climate change is not the only serious environmental issue facing the electricity industry.

Thankfully, lawmakers and electric utilities can implement six changes that would do much to avoid water shortages in the future. They could increase funding for research and development on new and alternative power plant cooling cycles, and place a moratorium on all new thermoelectric capacity additions, preserving the water available for other industrial, residential, and agricultural needs. They could aggressively implement energy efficiency and demand-side management programs to displace the need to construct new power plants. They could seriously deploy wind and solar power systems that use virtually no water when they generate electricity. (The combination of investments in energy efficiency, wind, and solar would not need to replace existing thermoelectric power plants, just offset new thermoelectric capacity additions.) Regulators could improve the feedback and information given to their electricity consumers about the electricity-water nexus and the water intensity of their own consumption habits. They could also form an interstate organisation or approach the federal government for assistance in designating National Electricity- Water Crisis Areas.

Interestingly, many of these actions would partially respond to a host of other challenges facing the industry unrelated to the use of water. The promotion of energy efficiency, wind, and solar would significantly reduce greenhouse gas emissions within the electricity sector, as all three technologies are less carbon intensive. A moratorium on future thermoelectric capacity additions would help diversify the fuels used to generate electricity, integrating more efficiency, solar, and wind into utility portfolios and serving as a hedge against fuel shortages and price volatility. Altering electricity prices would cut peak demand and create more informed consumers, an essential step towards any transition to a more decentralised electric utility sector involving small-scale distributed generation technologies. In this way, the proposals outlined here would not just save water and conserve electricity, but would also fight climate change, improve energy security, and educate electricity users. The benefits of such action surely extend beyond merely water and the electricity industry to society as a whole.

4

Anthropology and Policy in Global Context

Just three days after the disbanding of the summit, anthropologists gathered at the University of Georgia's Botanic Gardens for two days of presentations at the Environment, Resources, and Sustainability Conference to discuss policy issues for the 21st century. Each participant presented work in the areas identified as priorities at the Johannesburg summit.

However, no connection was made between our research and the United Nations debates. Consider what could be done with a systematic approach connecting the conference discussions to the global declarations of development need. An event of such high-level recognition as the Johannesburg summit could leverage bank and industry contributions and reinvigorate research partnerships among nonprofits, government, and educational institutions. Monitoring by researchers of policy events like the summit is critical to dissemination of research findings and formal input to future policy.

The structure of the Athens conference included five working groups: conservation management and local knowledge, environmental justice, consumption and globalization, fisheries, and agriculture and water. Subgroups concerned with each of these areas attempted summary and prescriptive statements aimed at policy formulations toward sustainable futures.

Within the agriculture and water group, subthemes were: 1) contested uses of water; 2) maintaining clean water while sustaining agricultural and other uses; 3) allocation of access to water resources among users; and 4) management plans that honor the needs and perspectives of all users. All the discussions were profoundly anthropological in the sense that they began with individuals and their communities and sought ways to bring these viewpoints to the table in policy formation. At the end of the two days we had not just narrated the details of field projects and administrative struggles. We observed that as each of us argued for placing these projects on the policy agendas of relevant agencies or corporations, the term "policy" took on some of that magical quality that surrounds other iconic words in our discipline: "culture," "mental model," "rites of passage," and "ethnicity." But though we think that understanding culture grants us influence over decision making, most of us have little experience in policy work, particularly when it comes to multinational situations. However poignantly we researchers articulate hardships, injustices, or inhumane conditions and place them in holistic cultural contexts, it is the rare researcher who has the opportunity to take the next step: making findings valuable to those who could change the conditions that affect people and communities. The way to give life to research is to see it in action, not just in debate. Agencies, nongovernmental organizations (NGOs), and corporations typically have not looked to anthropologists for guidance in the formation of policy unless they have anthropologists on staff.

It was around these challenges that much discussion centred at the conference. Considering the global context of the Johannesburg conference, perhaps there is cause for some optimism that a bottom-up approach might be more successful in reaching goals of the many organizations intent on addressing water issues throughout the world.

The Research-policy Cycle

To be considered for funding or included in policy discussions, usually, issues that researchers identify must be recognized by some politically and financially endowed body as worthy of placing on a policy agenda. The Johannesburg summit illustrated the cycle of policy formation and practice.

The first Summit on Sustainable Development in Rio de Janeiro in 1992 (the "Earth Summit") and the second in 2002 in Johannesburg reveal the high level recognition of global problems: conditions affecting one nation, or people, have direct and indirect impacts beyond immediate neighbours. For example, from the U.S. diplomatic perspective there is grave concern about political unrest fostered by unsustainable environmental conditions. Hence, the biggest budget contribution to sustainability projects is made through the U.S. Department of State.

A summit is not only an occasion for collaborative work among parties who otherwise work independently, it is also a means to name and institutionalize or politicize a problem. In the wake of these highly visible meetings and working groups, policy agendas have begun to emerge. Policy agendas are found within the United Nations, within and between nations, within NGOs, and also in a number of partnership agreements among private and public sector members as well as the NGOs. The U.S. Department of State Clean Water For People Initiative and the investment of $400 million by the United States to enhance water management worldwide are two examples of a high-level expression and support of policy. Once such policies with committed economic and political endowment exist, we can address anthropological involvement. What binds all of these agendas in principle is the political assertion by the United Nations that water is a high priority concern that must be addressed within and between nations for the physical health of the world's peoples, for peaceful relations among nations, and for the sustainability of the planet. Against that background an anthropological understanding of a community problem and the ability to recognize the genius of local systems of resource management are valuable.

Before we examine the contributions of the conference participants, we turn to two recent volumes, each focused on water and global issues, to delineate more precisely the predominant approaches expressed here. In their conclusion to a collected work on water access and use and power relationships, Donahue and Johnston (1998b:339-46) posit four dominant themes addressed by contributors to the volume. They order these as follows:

1. "Conflicts over rights and resources" in which power imbalances predominate. Local people's concerns are opposed to corporate and state agendas and practices; significant issues include local knowledge and sustainability.
2. "Project culture" that focuses on bureaucratic conceptualizations. This pits decision makers, usually official players, against those who are likely to experience the consequences of decisions and acts. This consideration also addresses "bureaucratic inertia," and interagency conflict, which is often the converse of meaningful action.
3. Development versus local sustainability of biological and cultural systems.
4. Crises with regard to water scarcity and quality, seen as cultural, social, and political issues.

The synopses presented here demonstrate the contributions of anthropological research and practice relating to the above thematic areas.

A second volume focuses on water issues relating to public health seen within a global context (Whiteford and Whiteford 2005). In their introductory statement, Linda and Scott Whiteford (2005:4) note that one of the "major threats to world stability" is "access to resources such as water." Anthropologists intent upon entering the discourse around this issue and becoming involved in practice addressing the crises, the need to engage in collaborative, interdisciplinary approaches, expanding our units of analysis to reflect the global milieu within which these crises materialize. This means taking account in our work of "international discourses, power structures, and commodity chains."

As Whiteford and Whiteford point out, international agencies are already significantly exploring approaches that have global reach. They note the implications, especially for local populations, of World Bank and World Health Organization policies that have come to privilege ideological perspectives like privatization. The oppositional response to some of these initiatives by local vulnerable populations has become the stuff of anthropological accounts, some of which are highlighted in this chapter.

Finally, with respect to anthropologists contributing to policy, Whiteford and Whiteford (2005:7) state: "the critical, structural approaches often employed in anthropology recognize that the social and political causes of structural violence against people based on gender, ethnicity, and economics require insertion into policy analysis." The contributions that anthropology can make to policy formulation may be obvious to us, but the enactment, as we see in the following cases, may be a more daunting task.

Anthropology and the Research-policy Cycle

In the Sugar Creek Project directed by anthropologist Richard Moore (2002) of Ohio State University we find an illustration of the research-policy cycle. In this case a problem was identified by a federal agency report, thereby gaining local and national visibility: The Environmental Protection Agency (EPA) reported environmental conditions harmful to ecosystems as well as to human communities. Sugar Creek was ranked the second most impaired watershed in the state of Ohio, just behind the Cuyahoga River in Cleveland, which burned in 1969, just four years before the creation of the Clean Water Act in 1973. EPA has developed the Total Maximum Daily Load (TMDL) plans for all impaired U.S. watersheds to reduce nonpoint-source pollution. The goal is for these TMDLs to be met by the year 2015.

Using the approach of participatory stakeholder involvement, whereby stakeholders play an active role in decision making, Moore created a research and action team that integrates stakeholders, specifically local farmers, with the interdisciplinary university research team. The research team consists of two anthropologists, a political scientist, an entomologist, a Geographical Information Systems (GIS) specialist, two water quality lab technicians, a Landsat specialist (NASA land satellite system), two natural-resources faculty specializing on riparian zones and microinvertibrates, a veterinary scientist specializing in E. coli pathogen transport, a USDA Agricultural Research

Service hydrologist, two EPA research scientists, and one farmer. That team is part of a larger group of researchers that focuses on sustainable agriculture. The research team works together with 10 to 12 conventional farmers who came together to learn how to reduce nonpoint-source pollution, a reaction to Sugar Creek's TMDL top priority label. Moore's team also works with two other farm teams, which have addressed two other subwatersheds of Sugar Creek. In 2003 the team started working with people on local tributaries within the watershed, a major break from traditional agency policy.

Hackenberg (2002) recommends that projects identified for policy involvement possess three key criteria:

1. The project is being exploited within the private sector.
2. Proposals are prepared as participatory interventions.
3. Policies should engage social units capable of collective action to both produce and distribute benefits, which will provide improvements in quality of life as defined by the participants.

Moore's Watershed Project Meets These Three Criteria

Field anthropologists engaging local farmers and other users of water regarding solutions to problems of allocation and supply may be cast in a different light from government policy makers, development economists, and functionaries representing international agencies. Often the anthropologist, operating through the lens of the village observer, may be criticized as simply myopic, harboring a perspective focused on the particular and incapable of grasping the larger picture. However, anthropology is positioned to identify problems by taking into consideration not only myriad factors that demand attention, but also the genius inherent in local systems of resource allocation and use where recognition is given to sustaining social and cultural orders.

Drawing on research in a number of local contexts, David Guillett addresses the new orthodoxy of demand allocation as a solution to an impending water crisis. From a point of departure suggested by Theodore Schulz and Robert Netting, he examines the institution of farmer managed irrigation systems (FMIS), nearly ubiquitous mechanisms for allocating water that take account of local conditions and imply sustainability. Guillett poses FMIS solutions against the market-based solutions increasingly promoted by international donor agencies, often as conditional contingencies of aid packages.

Contested Interests and Their Resolution

Anthropologists have long done studies of small-scale farming in a variety of environments and under different regimes of land ownership, usage, and transfer. Fewer studies have been carried out with large-scale corporate agriculture or collective farming, although a classic study of singular importance was undertaken by Walter Goldschmidt in the mid-1940s (Goldschmidt 1978). In addition to agricultural uses of water, papers considered a number of other topics, one of which involved the conflicting interests among various categories of users. Two of the papers dealt with contested issues between conservation and agriculture use and the attempts to mediate these.

In the first of these, a planning process for uses in a small watershed area of southern Illinois was instituted. At a later date a team of researchers from Southern Illinois University examined the planning process. Their findings indicated that the process lacked local legitimacy and failed to be effective as a democratic form of resource management because it did not involve a widely representative body in the planning and it failed to establish institutions for local management of the watershed.

A second instance concerns the planning process and eventual management of a newly created national monument in Montana. In this case Douglas Midgett, interested in approaches to dispute resolution in conflict situations in the rural American West, followed the process through its early stages, paying attention to how local viewpoints were accommodated and stakeholders saw their interests represented in the development of the monument management plan. Midgett participated in the "scoping" process contributing to the formulation of the plan and has interviewed and advised both local parties and Bureau of Land Management (BLM) personnel, who will construct the final plan and manage the monument.

Anthropologists as Participants in the Policy Arena

Diane Austin presented an example of an anthropological contribution to policy formulation at the University of Arizona where researchers and campus activists addressed water conservation and water use on campus. The work required "a thorough understanding of what is at stake for all interested parties as well as knowledge of the channels of communication, the social networks, and the issues under investigation." (Austin 2002:15). To make this expertise meaningful, she asserts that "we must examine and challenge our own biases and institutional policies that favour volume over substance, publication over application, exoticism over the familiar, and funding over need in project selection." (ibid., her emphasis)

Water resources and use are perennial concerns in the American Southwest. Although J. W. Powell recognized in the 19th century that aridity was the most important factor in human settlement in the West, boosters, planners, and policy makers have often avoided confronting this fact. A concerted conservation orientation has been shelved in favour of technological fixes, dam and reservoir construction, aqueducts that transport water hundreds of miles, and paper schemes that imagine even more fanciful "solutions." Within the Santa Cruz watershed that spans the border between southern Arizona and northern Sonora, Mexico, lie two major urban centres, Tucson and Ambos Nogales, the twin cities on either side of the border. Both urban areas depend for at least some of their water resources on the watershed, in-ground water, and limited surface flow. In examining two cases, Austin views attempts at a water conservation effort on the Arizona campus and describes a revegetation project in Nogales, Sonora, where forest destruction has contributed to desiccation and air-quality problems.

In this section we examine cases where anthropologists have become involved in policy issues at various stages, from advising and design to implementation on the ground, including direction of projects. We first look at a case where the anthropologist acted as a broker between conflicting interests and, in the effort to mediate the contentious issues, brought to the discussion an approach involving a proven dispute resolution procedure. In this case Inga Treitler was employed by the Oak Ridge National Laboratory to bring an anthropological imagination to the conversation, mediating between the Lummi Nation and hydroelectric development interests by illuminating cultural considerations that the Lummi people regarded as necessary to the discussion of environmental impacts.

Her examination highlighted the cultural dissonances between the Lummi view of the resources, the position taken by the Federal Energy Regulatory Commission, and the inability of the environmental impact statement process to resolve these differences. A second case involved anthropologists working in a team to undertake evaluation of government-sponsored water policy initiatives. The team approach to examine the effects of water policy reforms in Zimbabwe included the anthropological contribution of William Derman and Anne Ferguson. Because a focus of the study was local access to water resources and empowerment, it was deemed necessary to locate the local within contexts of national policymaking and the agendas of international donor agencies. Applied aspects of the research centred on bridging efforts to link various levels in collaborative work. In this instance the issue of "project culture," alluded to above, became a significant consideration in how those policy reforms were carried out.

Anthropological Research: Policy Implications

Finally, we present two contributions where the research of anthropologists, working alone, addressed issues having significant implications for policy. In one case the issue was local in scope, although it may have general application to a number of cases involving small-scale water-power generation. In the second, the issue was recognized as a fundamental problem of human exploitation of common resources, one encountered in a variety of historical and contemporary situations.

Richard Peterson's research addressed the cultural and social factors that influenced the successful implementation of development initiatives, in this case small-scale hydropower generators in tropical Africa. The natural resource in the form of permanent stream flows was abundant; the need for electricity to power machinery necessary to promote local productive activities was evident; and the projects seemed to fulfil development goals aimed at local enhancement and sustainability.

However, as the research demonstrates, concepts of common property may collide with perceived individual ownership, a problem that can be overcome only through understandings of cultural factors, including powerful mythic figures whose agency in human affairs is taken for granted. In the Peruvian Andes, Paul Trawick examined irrigation systems long in use by agricultural communities. Again, communal principles with regard to water allocation were in operation. Trawick's analysis of these situations took a more nuanced approach to the "tragedy of the commons" discourse, for he informed us of the local structures that preclude the dire consequences of abuses of common property. In situations of water scarcity he argued that schemes for determining access must recognize the principles and privilege local ingenuity, especially when market based allocation is contemplated as an alternative.

Water as Local Resource, Water as Global Resource

Although the issues examined in the works presented here illustrate the myriad problems emerging from the human uses of water, these by no means exhaust the list of issues that might concern anthropologists. When human manipulation of the land and its water resources is viewed in its entirety, this truism becomes abundantly clear. For example, none of the participants in the conference dealt with ground water or the problems of shrinking aquifers and implications for agricultural systems that depend upon these resources. Nor was the construction, and more recently, the removal of dams a topic our discussion. We know, however, of the dependence that states and communities have developed and of the disputes related to downstream flows that have surrounded discussions of the Missouri River system.

The scarcity of fresh water for human use worldwide has been tallied often: less than 1 percent in rivers, lakes, aquifers, plants, atmosphere, etc.; only one-hundredth of that available for drinking; and 40 percent of the world's population facing shortages daily (Johnston and Donahue 1998a:l). When we consider the other organisms with which we share the planet, the situation is equally dire: "fresh water ecosystems account for less than one percent of all habitat area on earth" (Postel and Richter 2003:26). Given this, we must approach the problems of water with the recognition that human beings are primarily responsible for the ecological degradation that has come to affect water quality and availability, and that it is only through human agency that these problems can be addressed. In the epilogue to their volume on rivers and management, Postel and Richter (2003:202-204) suggest three actions that must be undertaken if we are to manage the resource in a sustainable fashion:

- Human communities must integrate better with natural cycles.
- We must reduce pressures on nature through reduced population growth and reduced consumption.
- We must address the implications of the worldwide trend toward commoditization and privatization.

Many of the contributions included here touch on one or more of these issues, indicating with careful study of local cases the validity of these prescriptions. In the conclusion to their edited volume, Donahue and Johnston (1998:339-341) discuss how the intersection of organizational forms produces conflict as competing interests, and, perhaps more importantly, different cultural concepts of water collide. We have seen this in a number of our contributions, and some of the papers discuss at length processes of dispute resolution and mitigation that have had mixed results.

One observation that seems inescapable is that anthropologists can contribute both to the compilation of knowledge through their research and to the formation of policy related to water management. Although we cannot blind ourselves to global forces that impinge on the communities and people we study, we need nonetheless to maintain a focus on the local and to assert the validity of our approach. Nitin Desai (2001:45), addressing the International Conference on Freshwater in Bonn, Germany, in 2001, invokes his "three Ps"-political will, practical steps, and partnership-a programmatic approach that aims to involve human beings at various levels in pragmatic and cooperative efforts to deal with these problems. And, as we noted at the outset of this chapter, Desai envisions the critical task as getting it right at the village level. If policy begins here, there is clearly a role for anthropology.

New market-based management strategies may prompt conservation and lead to higher-value uses of water resources. Freshwater is a scarce and often threatened resource throughout much of the United States, but particularly in the arid West. Supplies are being depleted or degraded by unsustainable rates of groundwater use, contamination, and damage to aquatic ecosystems. Meanwhile, demands for water are rising with an expanding population, higher incomes, and a growing appreciation for the services and amenities provided by streams, lakes, and other aquatic resources. But the options for increasing supplies are expensive relative to current water prices and are often environmentally damaging. Providing for increasing water demands requires changes in how water has traditionally been managed and allocated among competing uses. There is a growing consensus that greater reliance on economic principles in managing and allocating water is critical for more efficient and sustainable use. More than a quarter century ago, the U.S. National Water Commission's Final Report to the President and to the Congress made a strong case for facilitating voluntary water transfers to promote a more-efficient allocation of scarce water resources and to curb the perceived need for additional water supply projects. In 1992, the International Conference on Water and the Environment in Dublin and the Earth Summit in Rio both endorsed viewing water as an economic good.

Introducing economic incentives was also one of the core recommendations of the World Bank policy paper on Water Resource Management prepared that same year. In 2000, the World Commission on Water for the 21st Century concluded that we are on a path toward a water crisis and that business as usual is unsustainable. The commission's proposals for changing course include recognizing that water is a scarce resource and that we need to manage it accordingly.

Water Markets

Markets and prices play a role in allocation of resources among competing uses, and they provide incentives to conserve and invest in new supplies. In a competitive economy, price adjustments and market transfers keep supply and demand in balance. Prices rise when demand increases faster than supply. Higher prices provide incentives to use less, to produce more, and to develop and adopt technologies that conserve use and increase output. Markets enable resources to move from lower to higher-value uses as conditions change. For example, water traditionally used for irrigation may be more valuable as a municipal water source as the demands of a nearby urban centre increase. Tradable water rights potentially can encourage conservation and a more economically efficient allocation of scarce water resources. Currently, water is underpriced and often allocated based on institutions established when water was not considered to be a scarce resource. Users pay nothing for the water itself. Municipal and industrial users typically pay a fee reflecting the costs of storing, delivering, and treating water supplies.

But even these costs are likely to be subsidized for irrigation, which commonly represents a region's largest water use. Without an opportunity to sell unused supplies, irrigators have little incentive to conserve water. With the introduction of tradable water rights, however, users value water in terms of its opportunity cost—the value they could get by selling water—rather than at the subsidized price they pay for it. In spite of their potential benefits and growing popularity, market forces have been slow to adapt to the reality of water scarcity. Efficient markets require well-defined, transferable property rights, and the full costs and benefits of a transfer must be borne by the buyers and sellers. Both the nature of the resource and the institutions that manage and allocate water can make it difficult to meet these conditions.

Obstacles

The variability of water supplies in time and space creates problems for establishing clear property rights. Driven by energy from the sun, water constantly evaporates from seas, lakes, and streams or transpires from plants, entering the atmosphere, then returning to earth through precipitation. Precipitation that is not quickly evaporated or transpired back to the atmosphere is the source of a region's renewable water supplies. This water flows into lakes, rivers, groundwater reservoirs, and eventually the ocean, unless it is first withdrawn for use.

Three basic systems—riparian rights, prior appropriation rights, and public permits—have developed for establishing rights to this water. Riparian rights. The common law system of riparian rights gives owners of the lands bordering a water body use of the water in ways that do not unduly inconvenience other riparian owners. Riparian rights have origins in the earliest legal systems establishing private ownership of land.

They continue to provide the basis of water law in many areas, including some arid Moslem and humid European countries and the eastern United States. These rights are poorly defined because shortages are shared by all riparian owners and use is subject to regulatory or judicial interpretation as to what is reasonable or might unduly inconvenience others. Moreover, riparian rights are not directly marketable because they are attached to the land and use is restricted to those lands.

Appropriative Rights

Constraints on transferring water to non-riparian lands and uncertainties such as how much water a riparian owner can use are obstacles to applying riparian rights in areas where streams are fewer and flows are smaller and less reliable. Consequently, in the arid and semiarid western United States, riparian rights were abandoned in favour of water rights based on prior appropriation.

Prior appropriation rights, which have been adapted by all 17 western states, have three principal features:

- Water rights are established by withdrawing water from its natural source and putting it to a beneficial use, such as irrigation. Unlike the owner of riparian rights, the party appropriating water does not have to be a riparian landowner, and the water does not have to be used on riparian lands.

- During periods of shortage, water is allocated according to the principle of "first in time, first in right." Thus, junior appropriators receive no water until senior appropriators—those with the oldest rights—have received their full allotment.
- Failure to use water for some period of time results in loss of the right. This provision creates a use-it-or-lose-it incentive, encouraging withdrawal even when the water contributes little if any value to the user.

Appropriative rights encourage the depletion of waters in a stream. Until recently, such instream flows lacked protection under state water laws. Unlike riparian rights, appropriative rights can be transferable, but sales are commonly restricted as to how and where water can be used.

Public Permits

Riparian and appropriative water rights initially were acquired without state interference. But as supplies become scarcer, governments are assuming a more active role in controlling water use. Some form of a permit system now governs use of at least some water in virtually every country. In principle, water permits can be auctioned by governments and bought and sold in private markets. In practice, however, permits are usually free, and transfers are limited by the nature of the right as well as by the infrastructure available to store and transport water.

Transferring water from one place or use to another will commonly affect third parties—those other than the buyer and seller. When a farmer sells—in effect, transfers—water to a city, the economic base of the water-exporting community may decline. And when a transfer alters the quantity of water in a stream, other stream users are likely to be affected. Indeed, changes in flow affect the amenities and recreational opportunities that rivers and lakes provide, as well as the individuals who enjoy them. Generally, these water services are public goods that are not marketed, because the public cannot be excluded from freely enjoying them. Thus when marketing water, the private sector tends to ignore the impacts of water transfers on the public goods the waters produce. Similarly, polluters underinvest in waste reduction and treatment when the costs of using water bodies for waste disposal are borne by society rather than the individual polluter.

During the past three decades, water-related investments and legislation in the United States have been driven largely by a desire to protect the resource and the public benefits it provides. The resulting environmental legislation and regulations have contributed to uncertainties over water rights. On one side are the traditional users with rights established when water was treated as a free resource and environmental impacts were ignored. On the other side are the more-recently empowered stakeholders armed with legislation designed to protect and restore environmental and recreational uses. While domestic, industrial, and agricultural users may compete for the water diverted from streams and reservoirs, all three groups vie with environmentalists and recreationists over the amount that can be withdrawn. Conflicts also arise over the priority that dam operators give to flood control, water supplies, hydropower production, fish habitat, and recreational opportunities. Without markets and prices to guide allocation of water and guide dam and reservoir management, conflicts are often played out in the courts or administrative proceedings.

Groundwater initially was treated as a resource that landowners could capture at will. But groundwater is often a common property resource that flows from one property to another until captured for use. Pumping can adversely affect third parties. One party's pumping can reduce the water available to neighboring water users, forcing them to pump from greater depths and lowering their well yields. Third parties may also be harmed if groundwater use reduces surface flows, causes saltwater intrusion into an aquifer, or results in the collapse of lands above underground aquifers that have been at least partially drained. The emergence of such impacts, along with improved knowledge of the links between ground and surface water, has led to restrictions on groundwater use. In the western United States, for instance, most states have adopted some form of a permit system for groundwater. But even in states lacking a permit system—such as Texas, which continues to grant landowners unrestricted rights to pump groundwater without liability for damages inflicted on others—landowners overlying a defined aquifer may voluntarily form a conservation district to regulate wells.

Opportunities

The absence of markets and market-based prices to allocate scarce supplies and guide water managers has resulted in large differences in the value of water among alternative uses. For example, most of the water rights in the western United States are held by farmers and irrigation districts, which pay only the modest cost of having water delivered to their farms. As a result, large quantities of water are applied liberally to relatively low-value crops. In some cases, simply leaving more water in the river to provide hydropower, recreation, and fish and wildlife habitat might increase the total value of the water to society. Similarly, selling water to urban areas that otherwise would invest in costly and often environmentally damaging water-supply projects might boost water values by an order of magnitude or more in some cases. Large differences in the value of water among alternative uses provide powerful incentives to overcome obstacles to transfers. As a result, water marketing is becoming common and increasingly innovative in several countries. A variety of market arrangements have emerged in the western United States to accommodate and respond to short-term fluctuations in supply and demand stemming from climate variability or other factors. These include leases, options to purchase water during dry periods, and water banking. The temporary nature of such transfers blunts a principal third-party concern that a transfer will undermine the economic and social viability of the water-exporting area.

Farmers with senior appropriative rights who grow annual crops might profit by selling an option to use some of their water during a drought and leaving some or all of their fields fallow. Growers threatened with the loss of long-term investments in orchards, or cities facing rationing, are able to more than compensate these farmers for losses incurred from temporarily leaving some of their lands fallow. Such transfers among farmers within the same irrigation district are common and often relatively easy to arrange. But cities seeking to transfer water away from an irrigation district are likely to encounter greater institutional obstacles and financial costs.

Water Banks

Water banks represent one solution to these obstacles. Water banks are designed to facilitate water transfers in response to short-term changes in supply and demand conditions. They enable the owner of a permanent water right to sell all or part of one year's entitlement. Thus, the entitlement is, in effect, leased or rented but not permanently transferred. The primary objective of a water bank is to bring together those who want to purchase water with those who are interested in selling their entitlements.

The bank provides several key functions. It determines which water use entitlements may be banked and the quantity of water associated with each entitlement. It determines who can rent or buy water, and it sets the rental rules. The bank may also determine how much water can be transferred without injury to third parties. Once the rules are established, the bank operates like a broker, accepting valid water use entitlements for deposit and making them available to those hoping to obtain water at a lower cost than they would otherwise have to pay.

California

A temporary federal water bank administered by the Bureau of Reclamation and endowed with federal funds was established in California during the 1976-77 drought to provide for water transfers within the agricultural sector. Then, in 1991, California established an emergency water bank in the fifth year of a prolonged drought, after legislative attempts to promote private transfers produced few transactions. The bank, which reallocated water among willing buyers and sellers, operated in 1991, 1992, and 1994. Initially, the bank purchased water for $125 per acre-foot (about 10.1 cents per cubic meter) and sold it for $175 an acre-foot plus delivery charges. In some cases, the delivery charges exceeded the initial cost of the water. Purchases by the bank proceeded slowly until the state guaranteed sellers that their price would be adjusted upward to reflect subsequent seasonal price increases. Farmers willing to idle land or shift from diverting surface water to pumping groundwater were the principal sellers. The bank purchased 800,000 acre-feet (nearly 1 billion cubic meters) of water, based on early estimates of critical need before the buyers made firm commitments.

California incurred a sizable financial loss when the bank was able to resell only about half of the water. Unsold water was used primarily for carryover storage and for reducing saltwater intrusion into the delta of the Sacramento and San Joaquin Rivers. In subsequent years, the state required a signed contract with a buyer before committing to purchase water. The third-party impacts associated with these transfers are unknown, but they were probably insignificant compared with the benefits of moving water to higher-value uses during a period of severe drought.

Idaho

Idaho established the nation s first permanent water-banking program in 1979, and several western states are considering similar actions. Idaho's bank allows temporary or permanent transfers of water rights, but rules governing where and how the water can be used are restrictive. For instance, out-of-state transfers are prohibited and irrigators receive preference over all other users in purchasing or renting banked water.

Australia

In Australia, the state of Victoria, which facilitates transfers, defines water rights as explicit shares of stored, rather than delivered, water. Under this system of capacity sharing, decisions regarding reservoir releases are made by individual owners of the rights rather than by a central authority. Reservoir operators serve like bankers, making releases on request. The operators also keep track of each owner's balance on a continuous basis by adding inflows and deducting releases and losses from evaporation and seepage. Water users control the timing of their deliveries, and transfers can be made simply by having the operator make the appropriate debit and credit.

Banks can operate at any administrative level, ranging from multi-state to water districts to ditch companies. They can be designed to manage different types of water-use entitlements. And they can facilitate temporary water transfers by developing clear, well-defined rules and procedures that reduce transaction costs. Temporary water transfers, however, are not particularly effective for adapting to the long-term demand and supply shifts that result from population and income growth, urbanization, rising values for instream flows, groundwater depletion, and climate change. Indeed, at some point, as supply and demand conditions change, the historical allocation of water rights becomes inefficient enough to warrant a permanent transfer of rights.

Keeping Deserts Green

Transfers of permanent water rights are permitted, subject to review of third-party considerations, in all the western U.S. states. The process of resolving third-party impacts is often slow, costly, and contentious, however, and the outcome of a proposed transfer can be uncertain. Those orchestrating the proposed transfers face either the challenge of proving that a change will not harm others or the added cost of compensating the third parties who might be adversely affected by the transfer.

Ongoing efforts to meet the water demands of the rapidly growing coastal area of southern California illustrate the challenges of securing additional water in a region where supplies are already fully developed and allocated. The task is made more urgent and difficult because access to some of the region's traditional sources has been blocked. As a result of environmental concerns, for instance, Los Angeles has been forced to reduce the water it takes from Mono Lake and Owens Valley. The region is also losing rights to the unused entitlements of other states to water from the Colorado River.

The Imperial Irrigation District (IID), in the southeastern corner of California, owns senior rights to much of California's share of the Colorado River. To help meet the growing demands of the state's southern coastal region, the Metropolitan Water District of Southern California (MWD) agreed in 1989 to invest approximately $115 million, plus about $3 million annually for operation and maintenance, to conserve water in the irrigation district through such measures as lining canals to prevent water loss through seepage. In exchange, the MWD acquired the rights to about 100,000 acre-feet of conserved water per year. More recently, San Diego—which depends on, but has a low priority claim to, MWD water supplies—agreed to fund additional conservation investments in the IID in return for conserved water.

These agreements illustrate the opportunities and obstacles associated with using market forces to reallocate water. In these cases, water was transferred from relatively inefficient, low-value agricultural uses to higher-value urban uses; IID's agricultural base was preserved through conservation investments; and the interests of neighboring U.S. irrigation districts were protected. Yet concluding these apparent win-win arrangements required nearly five years of often contentious negotiations among the participants and interested third parties.

Moreover, agreement in this instance was facilitated because the participants were able to ignore the adverse third-party impacts in Mexico. Indeed, before the canals were lined, some of the water seeping out of them had helped recharge groundwater aquifers used by Mexican farmers just across the international border. These third-party impacts were ignored because the Mexicans had no recognized legal claim to the water. Efforts to arrange interstate sales of Colorado River water have been less successful. The 1922 Colorado River Compact among seven western states divided the river equally between the upper basin states (Colorado, New Mexico, Utah, and Wyoming) and the lower basin states (Arizona, California, and Nevada). The upper basin states have never fully used their entitlements. The MWD therefore has been able to take the unused water for free. California, however, is being required to cut back on its use of these surplus flows, and the upper basin states are seeking opportunities to benefit from their full entitlements.

The potential for mutually profitable transfers from the underused upper to the overused lower basin has stimulated several proposed sales. In the 1980s, the Galloway Group, a Colorado corporation with claims to 1.3 million acre-feet of Colorado River water, proposed constructing reservoirs to produce hydropower and store water for leasing to Arizona and southern California. San Diego paid $10,000 for an option to lease 300,000 to 500,000 acre-feet per year for 40 years, but the project died under a flood of unresolved legal issues.

Las Vegas—which already uses most of Nevada's entitlement to the Colorado River—is currently seeking more water. Meanwhile, it lacks rights to surplus flows, and depletion of its groundwater is causing subsidence within the city. Unused upper-basin entitlements to the Colorado River are a logical source of additional supply, and Utah appears to be a willing seller.

But consummating this or any other water transfer between the two basins could require renegotiation of the 1922 compact. Transfers among the lower basin states encounter fewer legal hurdles but, because of their similar hydrology, offer fewer economic benefits than do transfers between the upper and lower basins. In 1997, the U.S. Department of Interior issued a ruling designed to encourage and facilitate voluntary transactions among the three lower-basin states. Arizona has established a Water Banking Authority to purchase its own unused entitlements for storage in groundwater basins and possible sale to California and Nevada. But the opportunities for profitable water transfers among these three states—each of which is trying to meet the demands of rapidly growing metropolitan areas—pale in comparison to the potential benefits of transfers to the lower basin from the upper basin, with its large quantities of unused entitlements.

Colorado-big Thompson

The U.S. Bureau of Reclamation's Colorado-Big Thompson project has been cited as a prime example of efficient water marketing. The project involves a series of reservoirs to capture part of the flow of the Colorado River and its tributaries. An average of 230,000 acre-feet of water annually is transferred through a tunnel from the western slopes of the Rocky Mountains to the Northern Colorado Water Conservancy District in northeastern Colorado. Rights to proportional shares of this water are freely traded, unencumbered by third-party concerns, within the district.

Under western law, downstream users generally own rights to the return flows of upstream users, and transfers must take account of downstream impacts. But since Colorado-Big Thompson water originates in another basin, the district has retained ownership of the return flow of the diverted water. This arrangement does not eliminate third-party impacts, but it does eliminate the need to consider them in transfer decisions. In this case, the benefits of being able to transfer water readily among agricultural, municipal, and industrial users within the district are likely much greater than the costs of ignoring the third-party impacts. But limiting sales to within the conservancy district precludes even more-profitable transactions that might take place with buyers outside the Northern Colorado Water Conservancy District. For example, the right to an acre-foot of water in perpetuity has sold for $3,500 more in the neighboring Denver suburbs than in the conservancy district.

Chile Waters

Chile, which has embraced market economy principles for the last quarter century, has introduced measures to encourage water marketing. In 1981, the nation separated water rights from land ownership and made these rights freely transferable. Views differ as to the extent and benefits of the water markets that emerged in response to these changes. One view holds that Chile's water markets function effectively; water moves from lower to higher value uses, prices are responsive to temporary as well as longer-term scarcity, and trading is active.

A less-optimistic view holds that transfers of water rights separate from land ownership are uncommon, involve only a small percentage of users, and result in little actual reallocation of supplies. In particular, several factors are cited as inhibiting sales of water rights separate from land ownership. These include the inflexibility of the existing canal systems for distributing water, uncertainty as to who actually owns the water rights, a rural culture that believes water should not be bought and sold separately from land, and slow and erratic administrative procedures designed to protect third-parties from injury.

Sustainable Management

A consensus is growing that sustainable economic development depends on treating water as a scarce resource and using economic principles to guide its management and allocation. Water markets are a means of introducing these principles and allocating supplies in response to changing supply-and-demand conditions. But marketing water differs in important ways from the sales of most goods and services. The fugitive nature of the resource, the variety of services it provides, and interdependence among users limit the potential for efficient water marketing. Additional constraints result from the laws, regulations, and treaties that establish rights to water and limit how it can be used.

Water resources within a basin—precipitation, runoff, water in lakes and streams, and groundwater—are interrelated. Water users become increasingly interdependent as supplies become scarcer. Dams, reservoirs, canals, pumps, and levees make water availability less dependent on the vicissitudes of the hydrologic cycle and more dependent on human decisions. This infrastructure broadens the opportunities for allocating supplies and generates new demands on the resource. Reservoir operations affect a variety of water uses, such as flood control, hydropower production, recreation, fish and wildlife habitat, navigation, water quality, and domestic, industrial, and agricultural water supplies. Allocating reservoir capacity for one use affects other users within the hydrologic system. Drawing down a reservoir for flood control, for example, may reduce available supplies when they are most valued for irrigation, hydropower production, navigation, or recreation.

Managing these reservoirs in a way that best serves society is a daunting challenge. In smaller river valleys where water uses are limited, capacity sharing, as practiced in Victoria, Australia, is a way of defining marketable water rights. In a larger, more complex system, the interdependence among users is too great to ignore. Market allocation of water or reservoir capacity in such a system would be inefficient and chaotic. While markets are not a panacea for achieving efficient and sustainable water use, they can play an important role in achieving these goals under some circumstances. Water markets, whether formal or informal, have a long history of facilitating transfers. Chile and the western United States, two areas where marketing is most advanced, illustrate both the opportunities and limitations of water markets. Water marketing in these countries has largely involved transfers from relatively low-value, inefficient irrigation use to higher-value domestic and industrial uses. Moreover, the sales often provide incentives and funds for water-conserving investments to protect the economic base of the water-exporting community.

Geography and institutional factors, however, have restricted development of water markets. Moving water outside of its natural channels is costly and subject to economies of scale. Chile's rivers flow from the Andes to the ocean in a series of small, steep-gradient rivers separated by hills. Consequently, it is expensive to move water from one watershed to another or from downstream to upstream areas within the same basin. The frustrated efforts to sell water rights from the upper to the lower Colorado River basins illustrate how institutional factors can limit potentially profitable transfers even where the infrastructure is in place to move water at low cost. The successes and failures in transferring water emphasize that clearly defined, transferable rights are a necessary, but not sufficient, condition for market transactions. Once transferable rights are established, the most important challenge for creating efficient water markets is developing procedures for expeditiously and fairly handling third-party impacts. Unfortunately, these impacts are not always obvious or quantifiable. Both Chile and the United States allow transfers subject to consideration of these impacts. But the judicial and administrative procedures used to assess these impacts and compensate third parties often impose a high hurdle for prospective buyers.

In spite of the obstacles, the potential gains of transferring water for new uses are encouraging the development of water marketing in many areas. The incentives for voluntary water transfers are strong and will continue to grow as the resource becomes scarcer and the costs of providing water for traditional uses increase.

A Hot Future

Global warming would likely add to the potential benefits of water transfers. A warming world would alter the hydrologic system and increase the demand for water. The magnitude, timing, and even direction of climate-induced changes in a region's water supplies are uncertain. The costs of building dams, reservoirs, and canals in anticipation of these uncertain changes are high. But reexamining reservoir operating rules, relaxing constraints on water use, and developing institutions to encourage voluntary exchanges of water through markets would create a system more efficient and able to adapt to whatever the future might bring.

Will we run out of fresh water in the 21st century? The media highlights the parched lands, dry riverbeds and springs and falling groundwater tables across the world daily. Over a billion people living in developing countries without access to safe drinking water are facing economic and water poverty. Another real and troubling indicator is the rapid rate of aquatic habitat degradation and biodiversity loss in the last century. Projected changes in climate due to greenhouse gases invariably portray a future world that is much drier in the tropics—where over half the world's population lives—and suggest a global increase in floods and droughts.

Is a global water crisis already upon us? The answer to this question seems to depend on who you ask. On the one hand, active voices such as Sandra Postel, Peter Gleick, Vandana Shiva, Lester Brown and Paul Elrich, as well as leaders of major global organizations with an interest in water, have been warning of an impending global water catastrophe. On the other hand, the mainstream academic community involved in hydrology and water has largely ignored the topic. For example, a Google search for "water crisis" leads to almost 1 million hits, but the same search on Google Scholar yields approximately 4,000 hits as compared to over 1 million Google Scholar hits for "climate change." Malay of these articles focus on policy solutions, but do not necessarily explore the nature of the problem in-depth. Furthermore, the literature is largely non-American and contains references to much of the same work. Introducing "global water crisis" into a Google search reduces the number of hits by a factor of ten. In fact, the handful of scientists who do study this problem have divergent opinions as to whether and when the world will run out of water. A handful of scholars—particularly economists—go so far as to claim that a global water crisis does not exist or is, at best, overstated.

These scholars generally find that, on the whole, water access is improving worldwide and that with continued efficiency enhancements, the amount of water will continue to meet existing demands. Perhaps the way the global water crisis has been defined—whether the world will run out of freshwater—is the wrong way to look at the problem. While there are many scholars looking at the range of localized and specific water challenges that are occurring around the globe, it seems that the academic community has yet to find success in accurately characterizing the sum of their parts. In this chapter, we argue that there are three distinct water crises—or challenges, depending on who you ask-that have yet to be systematically connected by scholars. It is by looking at how these three challenges are interrelated that we can better articulate the global characteristics of water resource dilemmas and, ultimately, identify the global factors that can help solve these dilemmas.

Reorienting the Debate: Three Crises Rolled into One

Three types of water crises appear prominently in academic and professional discourse. First, there is the crisis of access to safe drinking water. This includes the inability to provide basic infrastructure to store, treat and deliver water supplies to a large part of the world's population. Second, there is the crisis of pollution that is analogous to climate change in that it relates to the impact of by-products of resource use. Third, there is the crisis of scarcity, or resource depletion, which is analogous to the fear of running out of oil. Now that we have defined three types of water crises, we can examine what we know about them, how they are linked, to what extent they are global problems and, finally, what are some possible solutions.

The Access Crisis

Many people equate the global component of a water crisis with the vast number of people worldwide whose economic productivity and social development is limited by access to safe drinking water. For instance, the World Health Organization, the World Bank Group Development Education Program, Global

Water and the Global Water Challenge draw attention to the fact that over 1 billion people lack access to safe drinking water. As a result, the United Nations Millennium Development Goals, the World Water Forum and other groups have rallied around a common metric for this issue by measuring the number of people with access to safe drinking water. For example, one of the key targets under the Millennium Development Goals is to "reduce by half the proportion of people without sustainable access to safe drinking water." Although these goals have been lauded as important policy directives, the international community has not yet made much progress in meeting them. Why is it so difficult to meet these goals? A vast body of literature points to the technical, institutional and financial challenges involved in developing the infrastructure and systems needed for water storage, supply and treatment. As this body of literature has discussed, poor countries may not have access to sufficient capital to build large-scale infrastructure like reservoirs, water treatment plants or delivery systems. If they have donors to supply the initial capital, they often do not have the means to repay these loans. Some developing regions that acquire the resources to build new infrastructure later discover that they cannot afford to maintain it. Other times, donors build water supply projects that are grossly mismatched with the needs of local communities.

Despite these challenges, research and practical experience have shown that the water access problem can be addressed—at least superficially—with existing knowledge and resources. For example, numerous low-cost technologies are now available to treat water quickly without large-scale infrastructure. Simple and cost-effective infrastructure, like rainwater harvesting systems, is readily available to many water-scarce communities. Additionally, creative financing mechanisms—like public-private partnerships—have been adopted by many local communities to pay for new water supply systems. These types of interventions—as well as successfully developed large-scale water supply infrastructure—have helped increase the number of people with nominal access to safe drinking water from 77 percent in 1990 to 83 percent in 2002. It is because of these improvements in recent decades that some scholars have argued that a global water crisis does not exist.

The Pollution Crisis

Intertwined with the access crisis is the water pollution crisis. For those 1.1 billion individuals who lack access to safe drinking water, "safe" is often the key word. While the infrastructure for water storage and access is often available, sometimes the water is contaminated by chemicals, microbes or other pollutants that render it non-potable. Yet, just as we have the know-how to develop water supply technologies, we also have the know-how to treat contaminated water. In the last century, tremendous research efforts have translated into an ability to treat wastewater and remove many of the most well-known chemicals of concern, including the ability to reuse or release this treated water into the environment. Technological advancements and tighter environmental regulations have created major progress in controlling the pollution of water from point sources, such as industries and municipalities. Although exotic or emerging chemicals continue to be a concern, their control is an active area of research. Therefore, where there is political will and available funds, pollution emitted from point sources is now under control. However, the political will and funds available to control point sources are still limited in many regions of the world. Many poor countries face similar challenges in developing the infrastructure to treat water as well as in supplying water.

An even more difficult pollution problem to solve is that of non-point source pollution, which results from diffuse sources, such as farms, or is caused by atmospheric deposition from industrial polluters. This form of pollution can have wide-ranging impacts, both on human use and on ecology, particularly through the accumulation of contaminants in water bodies and through the biological food chain. Examples of large-scale and cumulative ecological effects include hypoxia in the Gulf of Mexico, pfiesteria in the Chesapeake Bay and the decimation of the Ganges River dolphins. Historically, the effects of non-point source pollution have been easier to ignore than point sources because they affect humans less directly and visibly than sludge coming out of a pipe and directly polluting a drinking water source. Often, the cumulative impacts of non-point source pollution do not show up until they harm habitat and species living in downstream estuaries, bays and wetlands. These impacts cannot be ignored forever. As China has recently discovered, the extensive pollutants entering its waterways from factory waste, agricultural runoff and municipal sewage have had a tremendous impact on the quality of their aquaculture, causing decreased international confidence in their seafood markets.

Although it is possible to reduce pollution from diffuse sources, it is typically challenging and costly First, limiting non-point pollution requires substantial time and effort to figure out from whom and where the pollution is coming from, especially when the total amount of contaminants is high and when large volumes of water are moved during rainfall. In any given watershed, we may generally know that nitrogen and phosphorous entering a river is coming from upstream farms or mercury is coming from deposition produced by regional power plants. However, it is often quite difficult—without direct and costly monitoring—to know how much each polluter contributes in a given location. Thus, this type of pollution is more difficult to regulate and control than point sources. The United States has struggled for decades to enforce the Total Maximum Daily Load (TMDL) requirements under the 1972 Clean

Water Act. These requirements call for states and the federal government to identify sources of pollutants for each of the nation's waterways and to set acceptable limits for each pollutant.

One approach to dealing with challenges of identifying non-point source polluters is to require all industries that produce diffuse pollution to adopt technologies or practices that reduce the flow of pollutants into waterways. Many of these technologies are known and relatively simple to adopt. For example, all farmers could plant riparian buffers to filter nutrients or cover crops to reduce leaching. Alternatively, ranchers could protect streams using fencing to keep livestock fecal matter from entering waterways. The problem with these solutions is often political.

If we examine the experiences of some of the wealthiest regions in the world, like the Chesapeake Bay watershed—which encompasses Washington, DC and six U.S. states—large industry polluters often vigorously oppose such regulatory actions because they affect their bottom line. Enforcement or consensual action is often hampered by the fact that there is very limited data on the effectiveness of these best management practices, and a cost-benefit analysis is therefore difficult.

Moreover, unlike point sources of pollution—where it is possible to estimate the quantities of pollutants in a watershed and the benefits of reducing those pollutants—predicting the effects of non-point source pollution is challenging. For example, stochastic extreme meteorological events can lead to sporadically large "loadings" of non-point source contaminants, which are difficult to predict.

The cumulative impacts of a series of such events over a long period of time, and over large distances, can make it even more difficult to understand and estimate these pollutants. Mechanisms of non-point source pollution may also entail transport through multiple media. For example, this may include volatilization into the atmosphere, followed by deposition at a different location through rainfall, followed by the binding of pollutants into riverine or lake sediments. These mechanisms can operate from local to regional to global scales. Consequently, cause-effect analysis and monitoring for compliance or scientific analysis are made considerably more difficult.

Various incentives, such as low-cost loans or tax breaks to encourage voluntary pollution reduction measures among industries, can also fail. It is difficult to monitor and enforce non-point source pollution standards in large watersheds with multiple and diffuse polluters. Often, the opportunity looms for any one polluter to catch a free ride off the efforts of others.

Moreover, when non-point source pollution crosses political boundaries, upstream states have little incentive to control or treat pollution for the benefit of downstream states—unless downstream states have substantial political or economic bargaining power. Thus, non-point source pollution remains a significant challenge for technical, as well as socioeconomic and political, considerations.

The Scarcity Crisis

The third water crisis is one of scarcity. Scarcity refers to a situation when the water supply is inadequate in relation to the water demand for basic human and ecological necessities, including the production of food and other economic goods. Scarcity is arguably the principle component to the threefold water crisis because scarcity can drive—or at least exacerbate—both water access and water pollution. A community that has pumped all of its shallow groundwater dry will find it much more expensive to build deeper wells or to build trans-basin diversions to bring surface water in from another region. Additionally, when water supplies are depleted from a watercourse, the pollutants that may have accumulated there over time are likely to be more concentrated, thereby exacerbating the pollution crisis. This connection between water scarcity and pollution can then, in turn, lead to problems of water access. For example, municipal water supply systems can face increased treatment costs when reservoirs or instream flows are low because pollutant concentrations increase in the water they must treat. Water scarcity not only makes it more difficult to get adequate and clean water to meet human needs, but also harms aquatic habitats and species downstream.

With regard to access, invariably the most economical sources are developed first. Therefore, as scarcity increases, the level and reliability of access to water suffers unless water system budgets are increased. As such, we implicitly recognize that price changes can regulate water demand but that such changes are, in turn, a measure of the scarcity of the resource. One-third of the developing world is expected to confront severe water shortages in this century due to increasing population size and changing climatic conditions. Subsequently, not only will the poor and the under-represented (e.g., non-human species surviving off ecosystems) have to struggle to find adequate water resources, they will have difficulty accessing safe drinking water free from the various forms of pollution previously mentioned.

Of course, similar to water access and pollution, there are mechanisms to mitigate water scarcity problems. In most parts of the world, temporal water scarcity is dealt with by building storage facilities, such as large dams and reservoirs that can be filled when natural flows are abundant and tapped when natural flows are scarce. When communities lack the space or political will to build new dams and reservoirs, some have developed conjunctive water management programs that capture surface water and store it in groundwater basins through recharge basins or injection wells, which are then pumped for use during times of drought. Similarly, communities commonly address spatial scarcity by building aqueducts and canals to move water from a region that is water-rich to a region that is water-poor. However, even when communities have the financial and institutional resources to build and maintain such infrastructure, a long-term drought or increasing water demand can render the infrastructure insufficient.

The recent problems in the Southwest of the United States—a region that relies on Colorado River water to supply most of its growing population—is an example of this challenge. The main reservoirs in this area—Lake Mead and Lake Powell—dropped from being nearly full in 1999 to half-full in 2007, leaving states bickering about what actions they will take if long-term drought and growth continue. These states are now scrambling to find alternatives to make up for the likely losses in the basin they are so dependent upon. If structural solutions fail, migration can be the ultimate consequence of local water scarcity.

Some scholars and water managers argue that water markets or pricing schemes—especially in wealthy places like the U.S. Southwest—are an effective alternative for addressing water scarcity. In an economic system, the price of the commodity increases as scarcity increases, thereby regulating demand. However, pricing water is typically ineffective in modulating water supply and demand. First, there is political opposition to the creation of water markets because water is deemed a necessity.

If water goes to the highest valued users, less wealthy individuals and species may be short-changed. Thus, even where markets exist, they are often regulated with pricing caps or only operate within limited sectors (e.g., between a small number of farmers). The legal and institutional frameworks that govern water allocation and rights can also create significant transaction costs that can hinder the effectiveness of even small-scale water markets. Such challenges indicate that although solutions to the scarcity are well-known, implementing those solutions is not easy when one considers the political, social and economic costs associated with alternative solutions.

Clarifying the Global Dimensions of Water Scarcity

Given the difficulty of addressing water scarcity, access and pollution at even local scales, it is easy to understand why water crises are ubiquitous today. Yet, thus far, in characterizing the three water crises and their common solutions, we have not addressed any underlying global dimension of these three related problems. The global aspect of the climate issue is fairly obvious: Air pollution of individual countries translates into a pollution of the global commons that then impacts everyone in the future. However, we typically do not think of water as a global commons.

The possibility that North India may run out of groundwater in a decade leading to a collapse of agriculture in India is not viewed as a global problem. Likewise, the fact that the Yellow River no longer makes it to the sea, the fact that an aquifer in Long Island has been depleted and the three-hour daily walk for poor-quality drinking water in rural Ethiopia are all perceived and felt as local or regional problems.

The discussion of global water crises refers to the vast number of people around the globe facing these problems. In essence, the global crisis is viewed as a collection of local crises—whether they are related to access, pollution or scarcity—for which there is a global policy, imperative. We rarely address the global elements of these individual problems. However, looking at the scarcity crisis more closely reveals a critical global issue.

Linking Local and Global Dimensions of Water Supply and Demand

It is apparent to anyone who has studied or thought about the global hydrological cycle that local water availability is intimately tied to the global and regional climatic processes that control the disposition and movement of atmospheric and oceanic water. Thus, the climate is a direct bridge between local rainfall or water availability, and global processes. In an era of climate change awareness, this connection is now well-documented and disseminated through popular and lay media. However, the implications of this connection in terms of local and global water supply are not often obvious.

At the global scale, it is possible that the hydrologic cycle may accelerate as climate changes, implying that rainfall patterns, and hence water availability, may change both in space and time. Additionally, climate phenomena—such as the El Nino Southern Oscillation—lead to concurrent and persistent droughts in large areas of the world. This implies that these areas are likely to experience water scarcity at the same time. With population growth, the ability of local storage infrastructure to buffer the population from the impacts of drought decreases. Further, noting that agriculture—and more specifically grain production—is a major water consumer, mega-droughts that span much of the globe limit the ability to address regional water scarcity through food imports from other regions. This emphasizes the dimensions of potential global water scarcity.

Many of our early population centres were built in areas with easy access to fresh water sources. Past civilizations that did not have continuous access to freshwater, or that existed in drought prone regions, sometimes perished (e.g., the Anasazi people of the U.S. Southwest). However, the development of local water storage and distribution infrastructure has since allowed societies to develop resilience to these climatic aberrations of supply. As a result, we still do not view water crises as stemming from a global hydrologic cycle that is akin to a global commons.

The ability to trade food is a primary factor that has allowed human populations to occupy certain geographical areas with much higher density than would be possible if all food had to be produced from locally available water. It is the virtual import/export of water through food that effectively connects the local dimensions of water scarcity to a global dimension. It is estimated that 30 percent of all water in global food today comes from a country other than the one in which the food is consumed.

This fraction is anticipated to grow, meaning that global market forces will play a role in both the supply and demand for local water resources. As globalization makes food trade an implicit mechanism for reducing the impacts of local food and water scarcity, the vulnerability to water scarcity will eventually extend to a global scale. One consequence of this extension could be global agricultural price shocks. Another consequence could be new or increased competition for water at the local level among differing sectors—municipal, industrial and agricultural-competition that can play out in political debates, court disputes and/or conflicts.

If climate change and the associated human migration projections pan out as indicated by current modelling efforts, water surplus, low population density and low-intensity land use areas—such as Canada and Siberia—may emerge as population centres and the agricultural production centres of the 21st century. There is some recognition of the possibility of this trend. Yet, to our knowledge, no formal analyses exist on the effects that such changes will have on resource use—water, food, land or energy—and ultimately what the socioeconomic implications of such changing patterns in resource use entail.

The Role of Agriculture

Agriculture is the dominant water user on the planet, accounting for 70 percent of global water use, on average, and greater than 90 percent in and or semiarid regions. Agricultural water use efficiency is typically very low. Although efficiency rates vary by crop type, for many crops only 10 to 20 percent of the water supplied in either irrigated (not drip) or rain-fed agriculture is transpired by the plant. The rest is lost either by direct evaporation from the soil or in the water distribution network. Furthermore, most of the agriculture on the planet is rain-fed, which has higher evaporation rates and lower crop yields than irrigated agriculture. Thus, given that agriculture is the dominant water use and dramatic reductions in agricultural water use are technically possible without an impact on food production, it is an obvious target for meeting the challenge of water scarcity.

If an order of magnitude reduction in agricultural water use could be achieved, there would be no global water scarcity for the foreseeable future, at least not as a constraint on global carrying capacity for humans and other life. This has led to slogans like "more crop per drop" and work towards identifying technologies such as drip irrigation or genetic modification of crops so they will consume less water or can be grown in salty water. Efficient irrigation timing—through weather monitoring, appropriate use of nutrients and the use of weather or climate forecasts—combined with appropriate cultivars, can also reduce water usage.

However, practical progress towards significantly reducing agricultural water use has been made in very few places, like Israel, where agriculture has shifted to drip or recycled waste water use and to high cash value crops from subsistence agriculture or the production of cereals. For instance, drip irrigation accounts for only about 1 percent of all global irrigation, even though it can substantially reduce water use relative to flood irrigation, which is the most common practice. Access to technology is only a small part of the solution. The economics, politics and sociology of agriculture—as well as education and cultural adaptation—play a significant role in limiting change. Globally, food and agricultural product prices were steady or declined in real dollars over the last fifty years, and are only now starting to increase.

Gains in productivity ushered in by technology—the Green Revolution—were responsible in part. Attempts to protect the rural sector—through water, energy and fertilizer subsidies—and support prices for agricultural products play a large and as yet, incompletely understood role in impacting local and global investment in reducing agricultural water use. For instance, an American cotton farmer who has access to advanced production technologies can achieve much higher yields than a cotton farmer in a developing country. Additionally, the American farmer, when given free or highly subsidized water, can then sell that cotton at a low price. Since the United States is a relatively large producer, it influences the global market price, thereby forcing other countries to politically provide similar support mechanisms to their farmers. This leads to profligate water use in both locations. The recent price increases reflect progressive limits on land and water productivity, the increasing population and the diversion of agricultural to non-food products like biofuels.

Subsidies have a role in development and could be redirected toward incentives for water conservation. This transition requires some degree of international policy dialogue and concurrence to address the inherent collective-action problem of responding to signals of water scarcity. However, in practice, internal politics decrees a maintenance and proliferation of the status quo. Additional supports are added to protect the sector from losses due to a lack of water available either due to upstream abstractions or due to a climatic exigency. For instance, farmers in the Punjab in India are lobbying for state funds to dig wells deeper, since the regional groundwater table has declined in places by about 1 meter per year, to a depth of about 400 feet. Added to that, energy for pumping is provided to groundwater users for free or at a highly subsidized rate. The result is that groundwater—which constitutes a fossil reserve—is being mined in many regions throughout India, and worldwide, where farmers have access to cheap energy and the financial resources needed to pay for increasingly deeper wells.

Towards Agricultural Water use Efficiency

In countries such as India and China where population densities are high, especially in rural areas, agricultural water use already poses a stress on urban consumption. Major urban areas do not have the ability to provide drinking water on a sustained basis, even though they constitute a higher value use compared to agriculture and are able to pay for such use. For example, many major Indian cities face severe water shortages, often limiting public access to water from a few hours per week to a few hours per day. In theory, substantial volumes of water could be transferred from rural agricultural users to these urban sectors if greater efficiencies in agricultural use can be achieved. Much of the problem with such solutions in India is political. Providing highly subsidized rural water is a political norm that is difficult to challenge, even though the subsidies preferentially benefit a limited number of rich farmers and not the masses for whom these measures are intended. Improvements in the rural economy that facilitate agricultural water use efficiency are key for benefiting urban users and ecosystems as well.

One approach to achieving improvement in water use efficiency is to create a situation that leads to more revenue per drop. Increased financial resilience could then justify individual or group investment in a technology that facilitates a more efficient production of goods, while still assuring a high reliability of the water supply. Such resilience can further diminish the internal political pressure to support agricultural subsidies, driven in part by the need to protect farmers from financial loss. For example, higher revenue crop production has become available to some small farmers in India and China through the introduction of contract farming by national and multinational corporations.

Under contract farming, farmers are provided with all inputs and technological training to grow high-cash value crops that are exported or processed into goods and paid at the time of harvesting. The approach removes credit risk, market risk and technology risk from the farmer and provides incentives for higher yields from the same inputs. However, it requires a much higher technological and information base to execute through all stages of the product supply chain. This may lead to a transition toward higher efficiency production and a reevaluation of the time horizon over which the future value of the fossil groundwater resource or investments in improved surface water irrigation infrastructure are assessed. The corporation may be better placed to buffer the financial risks from which the farmer has been freed, and innovations on managing and sharing the climate risk through insurance and other channels are emerging. Finally, the corporation can look at the global marketplace for agricultural commodities and optimize what should be planted—where, when and how—considering market, labour and water constraints at local, regional or global levels.

Global Awareness and Global Solutions

In summary, there is a water crisis in many regions of the world, and the problems will progressively become global. There are three crises, but the preeminent crisis is one of water scarcity One of the dimensions of global water crises is already obvious to many; the problems of water access, water quality and scarcity are felt nearly everywhere, most prominently with the world's poor and under-represented. When we focus on the issue of scarcity—which can drive the other two crises—it is easier to identify the global nature of these crises as well as the connection between global and local forces. One of the key global elements relating to scarcity is the role of climate change. While we are becoming more attuned to the impacts of climate change on local water supplies, to a large extent we have mitigated this global influence by developing infrastructure and other forms of adaptive water management tools. The biggest global force that currently drives water scarcity is agricultural water use. The global market for food contributes to the flows of water resources—at least virtual water—around the globe, intersecting with local economic forces that support inefficient water use within this sector. We argue that the solutions to the scarcity challenge—and ultimately to part of the access and pollution challenges—require both local and global action at the policy and economic level to improve agricultural water use.

The few solutions we have discussed in this chapter require significant policy changes at the local, national and international level. To remove the masks on water scarcity signals, nations have to find ways to either change or overcome the powerful interest groups in the agricultural sector. The incentives for any one country to shift policies are quite low without substantial international efforts to ensure that other major agricultural nations will also participate. If new agricultural supply-chain practices are part of the solution basket, industries need to be aware of these opportunities and national and local governments will need to find ways to incentivize and protect access rights for those farmers and industries willing to undertake new business practices. The precursor to such types of changes is, arguably, heightened awareness of the nature of the problem. Such awareness can occur if a professionalized forum is available to provide informed debate and opportunities for learning. Such an analytical forum is largely a public goods problem, and one that requires collective action at the international level.

The benefit of international efforts to collectively investigate global water problems is evident from the impact of the International Panel on Climate Change (IPCC). One might ask, "Why is it that climate change has become a much greater scientific and social concern than water scarcity, access to safe drinking water, water pollution or resource depletion?" Arguably, this is because the climate change research community has come to agreement on the nature of the problem and on a metric with which to measure it. Global temperature increase over the last century is now well-documented, and in the last two decades a lot of scientific and political effort has been expended to connect this to greenhouse gas emissions, deforestation and other human-induced changes. Model-based scenarios and scientific intuition are used to project dire consequences for the future of the planet and its societies if greenhouse gas and temperature trends continue. A synergy between scientific inquiry and political action drives the climate change agenda further on both fronts as a global issue. Many significant uncertainties remain with climate change projections, even though the causative factors are agreed upon. A large number of these uncertainties pertain to the details of the hydrologic cycle—how water availability will change and how the distribution of water through the atmosphere and through vegetation will modify climate. So far, we have only seen the beginnings of the conclusive impacts of climate change, and the action agenda is to find solutions before this crisis overwhelms us and causes irreversible damage. Our best guess is that these impacts may occur over decades.

Addressing climate change requires policy action, but also a search for technical solutions to quickly substitute fossil fuels, or effective carbon capture and sequestration. Energy conservation and use-efficiency improvements are important but will not be sufficient to meet the challenge. These observations emerge after nearly two decades of intense global debate—one that has been fueled and supported by information generated through extensive international research efforts—as well as the synthesis and scientific consensus coming out of the Intergovernmental Panel on Climate Change. The debate continues, but at a relatively high level of maturity, and rapid technological evolution to address the anticipated problems is in the offing.

By contrast, the dimensions of the current and emerging water security challenge are not as well understood. Even when attention has been drawn toward water problems, it has been fragmented across the three types of crises we face today and targeted towards the local dimensions of the problems, which incidentally, are similar to the symptoms of global energy and climate change issues. This has prevented a focus on the preeminent scarcity crisis, which requires much more information generation and understanding. It is in many ways a more complex problem than $C[O_2]$-induced climate change, given that its global and local causes are more closely identified with social factors, local access and supply chains, and with many layers of common pool resource problems. Developing a single technical solution that will address the problem is at first difficult to recognize. However, the agricultural use-efficiency issue emerges quickly as a dominant issue. Given that at least some technologies are readily available for use reductions, the solution in this case may very well be to first achieve conservation and efficiency improvements using these technologies. Innovations in corporate farming offer some promise to improve rural livelihoods, while providing access to global markets for agricultural products and facilitating the reduction of climate, market, credit and labour risks through efficient global pooling of financial resources. New technologies to reclaim contaminated or saline water may also be an important part of the equation. This could involve using solar or wind technology as energy sources or coupling them to efficient greenhouse-based agriculture. Greenhouse-based agriculture—which has been widely used in Israel—can produce high crop yields while recapturing much of the water transpired by the plants and evaporated from the soil. The reuse of water can lead to substantial reductions in overall water use from the agricultural sector. Current pessimism as to whether we will have sufficient water to support life on Earth could be transformed into optimism, if we could formulate a path working toward these types of innovations.

The formation of a global roundtable as a focal point for analysing the causes and anticipated future conditions of water scarcity could help achieve what the Intergovernmental Panel on Climate Change has done for climate change. An interesting difference in this regard is that the political dimensions and social importance of the water issues are much better recognized at the outset than they were for the climate change problem at the initiation of the IPCC. On the other hand, the causal structure and scientific bases for climate change and its impacts, as well as the supporting databases, were perhaps much clearer for climate change than for the water scarcity problem. Consequently, a global roundtable would need to focus, from the beginning, on the collection and analysis of the wide range of information sources that are relevant to understanding and predicting the causal structure of local and global water scarcity. It is through such a process that both the scientific and policy communities can begin to understand and respond effectively to the global drivers of water crises, fostering tools and choices that could help leverage the diverse and important local solutions that are already available.

5

Irrigation Water Deliveries Versus Consumptive Use

Irrigation of agricultural crops accounts for 80% to 90% of water withdrawals in the semi-arid and arid western United States. Much of the water used for irrigation yields relatively low net returns, so initiatives to meet growing urban and environmental demands by using price incentives are increasingly being discussed and in some cases put into effect. Under the prior appropriation water law doctrine prevalent in the western United States, which relies on the legal ownership of water rights, farmers usually do not pay for irrigation water per se. Rather, they are charged for (some of) the costs of capturing and transporting the water to their farms as well as the operational costs of the irrigation company or district servicing them.

However, since the beginning of the 1990s, efforts have been made at both the state and federal levels to encourage agricultural water conservation using a range of measures including water pricing. One example is the Central Valley Project Improvement Act passed by Congress in 1992 that introduced a tiered pricing schedule to be implemented at the irrigation district level.

Another example is the policy adopted in 1996 by the Bureau of Reclamation of the U.S. Department of the Interior which encourages and, in some cases, requires irrigation districts receiving federal reclamation water to incorporate conservation price incentives as part of their conservation plans. A detailed handbook was prepared in 1997 to help agricultural water districts and irrigation organisations develop price incentives to achieve more efficient water use throughout the western United States.

In a river basin context, when analysing competing water demands involving the agricultural sector, it is useful to distinguish among three measures of water use: water withdrawals, deliveries, and consumptive use. Withdrawal measures the amount of water diverted from the surface or ground water source. Delivery refers to the amount of water delivered to the place of use, i.e., the farm. Withdrawal minus conveyance losses equals delivery. Consumptive use is the amount of water that is actually depleted-lost to the atmosphere from evaporation and transpiration from plant and soil surfaces, and embodied in plant products.

Figure: *District to assess the system of ponds and irrigation water delivery*

In field irrigation situations, delivery exceeds consumptive use for several reasons, mainly because of on-farm transit losses and field losses due to the imprecision of the water application practices. For example, in the case of irrigation with open ditch with siphons, in order to assure sufficient water reaches plants at the end of the field, an excess is applied at the beginning. Farmers also may not know the precise amount of irrigation water needed and apply more water than strictly necessary. Furthermore, in some areas, water in excess of consumptive use may be applied to carry salts below the crop root zone. The difference between withdrawals and consumptive use is called return flow. With consumptive use typically amounting to 40% to 60% of deliveries, return flows represent a relatively large portion of deliveries, and in many river basins constitute an important part of the downstream water supply. Water economists have long recognised the importance of considering the different measures of water use in their analyses. Bain, Caves, and Margolis (1966) draw attention to the problem "of placing any emphasis on gross, rather than net, demands for water, since the overall adequacy of water supplies depends on the net consumption occurring in any given use". Nevertheless, most research on irrigation water demand and its price-responsiveness has concentrated on delivery demands and paid little attention to consumptive use demands.

This focus is not surprising because delivery is usually the farmers' decision variable, and information on the consumptive use of irrigated crops (and, for that matter, return flows) has not readily been obtained. However, in light of increasing competition between water demands, water policy initiatives such as volumetric charges for irrigation water need to be examined not only with regard to their effect on deliveries, but also on consumptive use and return flows. Although price incentives involving higher costs of irrigation water may change producers' requests for water deliveries, we expect that the amount of water consumed would be relatively less affected.

This analysis develops and implements an approach to measuring the relative effects of hypothesised price incentives on deliveries and consumption of irrigation water. While analyses of the demand for irrigation water use have appeared in the literature since the early 1960s, the elasticity estimates and related policy recommendations differ widely. Some researchers found that farmers are very unresponsive to changes in the price of water.

Therefore, they commonly caution against the use of pricing policy to bring about reductions in deliveries, because even for relatively small reductions, large price increases would be necessary-with significant effects on agricultural income and wealth. Other studies indicate a more elastic demand, and their authors conclude that pricing policy would be an effective instrument since it would provide the necessary incentives for farmers to adjust to rising prices by using irrigation water more efficiently. Despite the importance of knowing the responsiveness of farmers to price changes, little systematic study has been carried out on the factors which cause these differing findings.

Several recent papers have used models incorporating water deliveries (or withdrawals) and consumptive use to show that for river basin planning purposes the responsiveness of consumptive use to policy initiatives is often equally or more relevant than is responsiveness of deliveries (or withdrawals). With a focus on the western United States, these studies examine the impact of a number of policy measures such as improved on-farm irrigation efficiency, limits on deliveries, and increases in delivery prices. Using a numerical example, Huffaker and Whittlesey (1995) illustrate that improvements in on-farm irrigation efficiency appear to conserve water by reducing withdrawals, but in reality redistribute water between river and aquifer and do not necessarily change consumptive use. Focusing on the impacts of limits in agricultural water deliveries, Bernardo et al. (1987) use a farm-level crop simulation/mathematical programming model to show that through better-timed irrigations large decreases in delivery demands may be attained with only marginal reductions in consumptive use and crop yields. Similarly, Burke et al. (2001) explore the effectiveness of limiting deliveries with the goal of conserving water for alternative uses. By linking an on-farm economic decision model with a basin-wide hydrologie model of return flow, they demonstrate that when substitution of technology for water supply is allowed for (in addition to crop switching and land fallowing), the resulting decrease in consumptive use is considerably less than the reduction in deliveries.

Huffaker et al. (1998) analyse the impact of price changes for irrigation deliveries on agricultural water conservation based on theoretical derivations. According to their findings, the common presumption that increasing prices for deliveries leads to agricultural water conservation at levels directly related to the price elasticity of delivery demand is valid only where water unconsumed by crops is irretrievably lost to the river basin.

In that special case, instream flows are reduced by the amount of water applied, and the reduction depends on the price elasticity of delivery demand. However, in the presence of return flows, instream flows decrease by the proportion of water applied which is consumptively used. Farmers would be encouraged by increasing block rates to reduce deliveries but, as improved irrigation technologies are adopted, they would also increase the efficiency with which the reduced deliveries are consumed in crop production. Thus the impact on instream flows is empirically uncertain. Huffaker et al. conclude this uncertainty can only be resolved by assessing water price in terms of consumptive use instead of water delivery.

The model we develop aims to address this issue raised by Huffaker et al. (1998) (and by Bain, Caves, and Margolis more than three decades earlier) by empirically analysing the effect of hypothetical price increases for irrigation water on both the demand for deliveries and the derived demand for consumptive use. Our estimates are based on a two-stage water-crop simulation/linear programming model that was applied to the New Cache La Poudre Irrigation Company (NCLPIC), one of the dozen or so major irrigator-owned cooperatives in the lower South Platte Basin near Greeley, Colorado. Flows from the South Platte River and its tributaries serve the major urban-industrial centres and the most important agricultural region of Colorado. Over 80% of the water withdrawals from the South Platte are used for irrigation.

A linear programming model was chosen to estimate delivery and consumptive use demands since it can be easily adapted to represent numerous options available to farmers for adjusting to increased water prices. To carefully reflect the yield effects of reduced deliveries, the linear programming model incorporates water-crop production functions computed with a transient-state crop simulation model. An innovative feature of the water-crop production functions used in this analysis is that they show yield and consumptive use not only as a function of the amount of water applied during the season, but also as a function of the number and timing of irrigation events. This allows us to estimate the responsiveness of both deliveries and consumptive use to price changes under a wide range of adjustment options, including changes in irrigated acreage, cropping mix, irrigation technology, and irrigation scheduling.

Based on the model findings, we estimate the price-responsiveness of the demand for deliveries and the derived demand for consumptive use as well as the effect on net farm income, and discuss resulting policy considerations. The quantitative estimates of price elasticities of consumptive use demand are, to the best of our knowledge, the first in the literature to date. A comparison between the price elasticities of delivery and consumptive use demands reveals that consumptive use demand tends to be significantly less responsive to increased water price than delivery demand.

This finding implies that price incentives for deliveries are likely to have only limited impacts on basinwide water consumption and would not make much additional water available for competing demands. We also examine the influence of different model assumptions regarding the substitutability of other resources for water on the estimated elasticities, and find that both delivery and consumptive use demand elasticities are quite sensitive to model specification.

Previous Analyses of Price-Responsiveness of Deliveries

Estimates of the shape of the delivery demand function are commonly based on the use of mathematical programming models, especially linear programming models. Gardner (1983) presents an overview of the studies carried out in California during the 1960s and 1970s. The early studies often intended to show that the delivery demand is more price-responsive than generally believed, and that even for very low prices it is not perfectly inelastic as the U.S. Bureau of Reclamation had claimed in the past. Later studies have constructed subregional or regional demand functions from models of representative farms, and commonly calculated responsiveness by either arc-elasticity estimates along the stepped demand curve or by fitting continuous regression equations to the parametric data.

The results typically report either an inelastic estimate for the whole price range considered, or an inelastic estimate for the lower prices and a less inelastic or elastic estimate for the higher prices. The shape of the demand function may be influenced by various factors considered in the model such as the quality of the soil, the length of the furrow, the distinction between surface and groundwater, the product price, and the elasticity of the product demand curve.

During the 1970s and early 1980s, estimates of delivery demands and their shape have been developed with statistical crop-water production functions based on data from field experiments conducted at state experiment stations. These analyses suggest optimal water applications per acre are very unresponsive to changes in price. A reason for this finding is that, while allowing changes in water applications for each of a few selected crops, these studies did not permit shifts in the cropping pattern or provide possibilities for substituting other inputs (e.g., labour or fertilizer) or alternative water application technologies for water.

More recently, econometric studies have used secondary data reflecting actual farmer behaviour. Their estimates tend to be more inelastic than suggested by the mathematical programming models, reflecting, at least in part, the differing assumptions and limitations of the two model types. Econometric models produce positive estimates based on historical data which often show little fluctuation in water prices, while mathematical programming models produce normative estimates based on both historical and synthetic data. The latter can be adapted to represent a wide range of scenarios, and model the impacts of policies for which no historical observations need to exist.

Overall, elasticity estimates vary widely-not only between model types, but also among mathematical programming models. Hartman and Whittlesey (1961), in an early study based on representative farms in Colorado, already noted that in addition to factors such as input and output prices, the kinds of adjustments farmers are allowed to make in the model in response to changes in water supply determine the value of additional water, and thus the shape of the demand curve. Our analysis builds on these findings and explores in more detail the effects of model formulation on the shape of both the delivery demand curve and the consumptive use demand curve.

Modelling Procedure

The model developed here to measure price-responsiveness of demand for deliveries and consumptive use consists of two parts, an agronomic and an economic model. The data are based on the irrigation practices of NCLPIC in Weld County, Colorado. NCLPIC has relatively senior water rights for river flow, but reservoir water and groundwater from the unconfined shallow alluvial aquifer along the South Platte River are also used for irrigation.

The major crops are corn grain, alfalfa, edible dry beans, corn silage, and sugar beets. Farmers almost exclusively use surface technologies for distributing water, including open ditches with siphons, and gated and flexible pipes with and without surge. They typically apply several irrigations per crop and season, each with a more or less fixed amount of water. With a lack of on-farm data on yield responses to different water supplies for the study area, water-crop production functions were computed using a transient-state crop simulation model. The model was applied to the five main crops grown in the service area of NCLPIC assuming an average-weather year. The main features and results of the crop simulation model are summarised. A detailed description including its input parameters is given by Scheierling, Cardon, and Young (1997).

While crop simulation models employed by economists typically treat the water input as a single value of water applied during the season, our model was adapted to capture the effects of irrigation timing as discrete-input events. The model outputs are watercrop production functions that show the impact of alternative irrigation schedules on consumptive use and yield. Each point represents a relationship between yield (and, implicitly, consumptive use) and the number and timing of irrigation water applications. Using such production functions as an input in the economic model has two advantages. First, it allows us to incorporate irrigation scheduling as a decision variable of farmers faced with increasing water prices; and second, in addition to the conventional relationship between water application and yield, it provides estimates for the policy-relevant variable under consideration i.e., the consumptive use associated with a given number and timing of irrigations.

A deterministic single-period linear programming model was developed to incorporate the physical relationships between water application and yield/ consumptive use derived with the crop simulation model. Formulated for the long run, the economic model computes the net return-maximising water applications in the study area in response to hypothesised alternative levels of water price, together with a derived demand for consumptive use. Changes in the output supply from the study area were not expected to be large enough to significantly affect regional product prices. An important assumption is that farmers are well informed not only about prices of inputs and outputs, but also about optimally timed irrigation schedules so that they apply limited irrigation water only in these combinations which result in the highest crop yields. Each crop can be irrigated with any of the five irrigation technologies. As in Caswell and Zilberman (1985), possible yield differences between technologies were not considered. These assumptions generated 245 activities.

Unit net returns were calculated for each activity based on the residual imputation approach (Young, 1996). These net returns are calculated as total revenue per acre (derived by multiplying the yield estimate from the simulation model by the crop price) minus variable costs (labour, materials, fuels, but exclusive of irrigation water costs) and annual overhead (including management) and annualised capital costs (inclusive of a land charge estimated at the value of the land in its next-best use, which is assumed to be the production of nonirrigated winter wheat). The resulting residual was imputed to the water resource and used as a value for net income in the objective function. Volatility and inflation were removed from crop prices by taking a five year average of prices for the period 1989-1993 deflated with the GNP Implicit Price Deflator (Colorado Agricultural Statistics Service). Variable costs were taken from crop budgets for the Nebraska Panhandle and adjusted to Weld County conditions based on the advice of Colorado State University extension agents. Alfalfa establishment costs were amortised over the average stand life in northeastern Colorado. Constraints in the model were defined for land (the service area of NCLPIC, amounting to about 40,000 acres) and water deliveries.

The latter include surface water (estimated based on average annual withdrawals from river flow and reservoir rights) and groundwater (estimated based on annual well allotments), and amount up to 120,324 acre-feet available in a typical year. An accounting constraint was formulated to measure consumptive use for each production activity. Other constraints were formulated to reflect the cropping pattern in the study area. Dry beans were limited to 17% of the total irrigated area, considering farmers' risk aversion as a result of highly variable bean prices. Sugar beets were allowed on no more than 7% of the area due to the contractual quotas imposed by the processor. Corn silage may be grown on up to 12% of the area, and alfalfa on up to 27% taking into account the magnitude of demand for fodder from nearby feedlots.

Empirical Results

Computations from the economic model focus on the impact of hypothesised price increases on irrigation water deliveries and consumptive use for the study area. Three scenarios with varying on-farm adjustment possibilities to changes in water price were analysed. Scenario 1 limits adjustments to changes in irrigated acreage and crop mix. It further assumes that the irrigation technology for the whole service area is open ditch with siphons, and the number of irrigations applied to each crop cannot be changed from the one which is optimal at the maximum water availability of 120,324 acre-feet. Growing crops with no irrigation is possible. Scenario 2 is similar to scenario I, except the number of irrigations can be varied and decreased down to zero. Scenario 3 allows for the widest range of adjustments including changes in irrigated acreage, the crop mix, the number of irrigations, and irrigation technologies.

As water price starts to rise from zero, the model indicates that deliveries are most quickly reduced in scenario 3, which allows for the most adjustment possibilities. Farmers in scenarios 1 and 2 with fewer adjustment options are initially much slower in reducing deliveries. As water prices reach very high levels, farmers in scenario I are predicted to be the first to stop irrigating, whereas farmers in scenario 3 continue to demand some irrigation water up to a water price of $292. In scenario 1, where adjustment possibilities are limited to changing irrigated acreage and cropping mix and switching to no irrigation, the model implies consumptive use would decline almost proportionally with deliveries.

But consumptive use values decrease only very gradually in scenario 3 because farmers can switch to irrigation technologies with higher application efficiencies without much effect on consumptive use (and yields). At least initially, they can also reduce the number of irrigations without much impact on consumptive use. The number of irrigations for alfalfa, for example, can be reduced to five, or even four, before consumptive use values and yields change significantly. For scenario 3, this stage is predicted to set in when deliveries fall below about 52,000 acre-feet. When prices are so high that deliveries cease, farmers in all scenarios are likely to continue to grow on part of the land crops such as alfalfa which have positive net returns for zero irrigations, and thus consumptive use does not drop to zero.

With an almost linear relationship between deliveries and consumptive use, scenario 1 exhibits a consumptive use demand which, over a wide price range, is similar in shape to the corresponding delivery demand. In contrast, consumptive use demands under scenarios 2 and 3 are predicted to diverge substantially from their respective delivery demands. Specifically, although farmers with more adjustment options respond to rising water prices by decreasing deliveries, their consumptive use values decrease at a much slower rate. At very high prices in all scenarios, consumptive use coming from rainfall and drawdown of soil moisture is estimated to remain at about 20,000 acre-feet. As water prices start to rise, net incomes in scenarios 1 and 2 fall relatively rapidly, while in scenario 3 changes in net income are initially much less dramatic. As water prices reach very high levels, farmers in all scenarios continue to earn a positive net income since they still have some crop production (and some consumptive use).

Under existing conditions in the lower South Platte Basin where irrigation water costs are quite low, water price increases in the lower ranges were considered to be the ones most relevant for examining the impact on deliveries and consumptive use implied by our analysis. The price changes included comprise an increase from zero to $30, and from $30 to $60 per acre-foot of water. Changes in deliveries and consumptive use, as well as in net income, were examined in both absolute and percentage terms. For a price increase to $30 per acre-foot, scenarios 1 and 2 show decreases in deliveries of slightly less than a third, while the more realistic scenario 3 exhibits a relatively large reduction of nearly 50%. However, for a price increase to $60, deliveries in scenarios 1 and 2 with less adjustment options are much more reduced (by 82% and 78%, respectively) than in scenario 3 where the reduction amounts to 69%.

The predicted changes in consumptive use resulting from the two price increases tend to differ from the changes in deliveries depending on the adjustment options of the particular model formulation. That is, the more adjustment options a scenario allows, the larger the difference between changes in deliveries and consumptive use tends to be. The predicted difference is especially pronounced at a water price increase from $30 to $60 per acre-foot. Under scenario 1, for example, deliveries would be reduced by 82%, while consumptive use would decline by 53%. Under scenario 3, which allows for investments in improved water use efficiency, the reduction in consumptive use would be overwhelmingly smaller than the reduction for deliveries (15% as compared to 69%).

Reductions in net income associated with increased price are estimated to be highest in scenario 1. At $30 per acre-foot, even though deliveries are only reduced by a third, net income decreases by more than half as a result of the pricing policy. Irrigators in scenario 3 still would receive 63% of their initial net income even though deliveries are reduced by half. At $60 per acre-foot, farmers in scenarios 1 and 2 are faced with a net income decrease of 79% and 70%, respectively, while farmers in scenario 3 still make 41% of their initial net income. This is because the potential to substitute capital for expensive water allows profitable production to persist in scenario 3 in contrast to the scenarios with more limited adjustment options.

To further illustrate the different effects of water price increases on deliveries and consumptive use, arc elasticities were calculated for delivery and consumptive use demands. The arc formula computes an elasticity at an average between two points, and allows for easy comparison between different scenarios. In the lowest price range (up to $30 per acre-foot), the demand for deliveries is more inelastic for scenarios 1 and 2 than for scenario 3 with more adjustment options, because initially these scenarios require relatively higher price increases for an adjustment, and thus a change in demand for deliveries, to take place. In the price range between $30 and $60 per acre-foot, the delivery demands for scenarios 1 and 2 become elastic, while the delivery demand for scenario 3 remains inelastic. As prices rise to higher levels, the model suggests that the delivery demands for all scenarios become elastic at some point. But especially for scenario 1, delivery demand tends to become more elastic faster because, as irrigation is ceased on more and more crops and irrigated acreage is given up, deliveries decrease more rapidly toward zero. (Switches to inelastic or perfectly inelastic estimates at some of the higher price ranges are caused by high "steps" in the demand curves, where deliveries are estimated to remain unchanged even though prices continue to rise.)

Consumptive use demand also tends to become less inelastic as water prices start to rise. But again, the predicted effects of increasing water prices are seen to depend on the scenario. The consumptive use demands for scenarios 2 and 3 which allow for more substitution possibilities as water cost increases-tend to be more inelastic than those of scenario 1. Further, consumptive use demands in all scenarios are more inelastic than their respective delivery demands over all price ranges. Thus the findings confirm that the price elasticity of consumptive use demand generally cannot be assumed to equal the price elasticity of delivery demand and, in particular, that the demand for consumptive use is likely to be much less price-responsive than the demand for deliveries.

The results reported provide evidence that elasticity estimates for both delivery demand and consumptive use demand are very dependent on the model framework within which they are derived. In the model formulations where farmers' adjustment options are relatively limited, there is a strong correlation between deliveries and consumptive use. Yet for the more realistically formulated scenarios, the correlation between deliveries and consumptive use is likely to be much weaker. This is because delivery is an input factor which, at least to some extent, can be substituted with other input factors such as management (adaptation in the number of irrigations) and/or capital (change to irrigation technologies with higher application efficiencies). Model formulations incorporating these substitutions indicate that they enable farmers to significantly reduce deliveries in response to price increases; at the same time, they can prevent large reductions in consumptive use, and agricultural production, over a relatively wide price range.

The elasticity estimates are also influenced by the method used to calculate them. In our case, most elasticity values would be different if the endpoints between which the arc elasticities are calculated were changed. Therefore, the emphasis here is not on the particular magnitudes of the elasticity estimates, but on the direction of their change depending on the model formulation, the price range considered, and the focus on either delivery or consumptive use demand. Our findings on the price elasticities of delivery demand are in line with previous results of linear programming models which indicated an inelastic demand for lower prices and a less inelastic demand for higher prices. However, they do not support the common presumption that in the case of an inelastic delivery demand, the use of pricing policy would not be very effective in bringing about reductions in deliveries-because, as the argument goes, even for relatively small reductions, large price increases would be required which in turn would cause large negative effects on income and wealth. Instead, this research suggests, especially for the more realistic scenarios with a range of adjustment options, an inelastic delivery demand does not necessarily imply that deliveries cannot be substantially reduced as the price rises. If in scenario 3 the price increases to $30 per acre-foot, the estimate of the delivery demand elasticity is -0.33 but deliveries would be reduced from 120,324 to 60,584 acre-feet. The modelling approach presented focuses on farmers' responses to hypothesised increases in irrigation water prices. To empirically examine basin-wide hydrologie and ecological effects of rising irrigation water prices and reduced agricultural deliveries (and withdrawals), a river basin optimisation model would have to be used such as, for example, Booker and Young (1994). Basin-wide effects are likely to include increased instream flows that could help restore river ecosystems. They could also involve improved water quality by leaving more native water in-stream for pollution dilution, and by reducing irrigation return flows with polluting salt, nutrient and pesticide loading.

Policy Considerations

Our empirical results have important implications for irrigation water pricing policies in river basins where water is withdrawn from surface diversions and/or shallow alluvial aquifers, and where return flows constitute a significant part of the downstream water supplies. This would be the case in Colorado's South Platte Basin where annual surface withdrawals are estimated to amount to about 2.5 times the annual native surface water flows, implying return flows are significant sources of downstream water supplies.

Based on results from the more realistic scenarios, even if delivery demands are price-inelastic, volumetric water charges could bring about large reductions in deliveries while not being very effective in reducing water consumption. They would also have the potential of altering basin hydrology, by reducing the magnitude and changing the timing of return flows. In the South Platte Basin, discharges from the alluvial aquifer (largely fed by irrigation return flows) increase late summer streamflows, which can help meet downstream irrigation demands for extended periods after peak runoff from snowmelt. The aquifer functions in the same manner as a reservoir, with irrigation withdrawals in excess of consumptive use contributing to storage and return flows representing releases. A water pricing policy that would significantly decrease deliveries, and return flows, could negatively impact water users who depend on these return flows downstream. Changes in the return flow regime could also impact groundwater levels and pumping costs. (Reduced return flows would not be a concern in the lower reaches of watersheds where there is no downstream dependence on return flows.)

If the goal is to make additional water available for alternative uses, in many river basins pricing of irrigation water deliveries would be a relatively ineffective policy instrument because, under realistic assumptions regarding potential farmer adjustments, net downstream river flows would not change much over a wide price range. In such hydrologie settings, the focus needs to be on measures that actually reduce consumptive use. Structural measures, such as land leveling or replacing open ditches with underground pipes, which mainly decrease transpiration from soil surfaces, would be one such approach. However, the amount of water involved is likely to be relatively small so that, without subsidies, the economic incentive for any single farmer to invest in such measures is minimal.

Larger quantities of water can be made available by changes in the crop mix, including switching to crops with lower seasonal consumptive use, or reducing irrigated acreage and switching to dryland crops. In the lower South Platte Basin, for example, farmers could switch from feed crops (alfalfa and corn) to vegetables such as carrots, which have a much lower water consumption and potentially generate much higher net income. However, these changes would involve making wholesale modifications in farm operations and entering a more dynamic marketing environment. Thus, they are not likely to occur on a widespread basis. To encourage such crop switches as well as potential transitions to rainfed agriculture or rangeland-which otherwise would only occur at very high water prices-subsidies or outright purchases of water rights would be necessary. Also, current federal subsidies for crops with high water consumption (such as sugar beets) might be reoriented to low-consumption/ dryland crops in order to support increased water use in nonagricultural sectors.

If volumetric charges were imposed on consumptive use rather than on delivery amounts, irrigation water pricing could theoretically be an effective policy instrument for reducing the amount of water consumed. Similarly, the problems surrounding irrigation return flows could be avoided if the surface and ground water rights under the prior appropriation doctrine were converted from withdrawal or delivery to consumptive use amounts. But measuring and monitoring actual consumptive use would still be difficult and costly. Short of these changes, policies regarding water transfers need to emphasize the importance of consumptive use rather than withdrawals and deliveries.

As water scarcity increases in the semi-arid West, it is often suggested that an appropriate pricing policy for irrigation water could reduce agricultural water use by providing stimuli for farmers to use water more efficiently, and make water available for higher valued nonagricultural uses. A number of earlier studies have noted that basin-wide water conservation depends less on changes in agricultural withdrawals and deliveries than on reductions in consumptive use. Our analysis extends this research by modelling both the demand for deliveries and the derived demand for consumptive use for a full range of hypothetical price changes.

Data from an irrigation company in Colorado's South Platte River basin were used as the basis for a water-crop simulation/linear programming approach designed to predict farmers' crop and water management responses to increasing water prices. In the context of the irrigation company, assuming an average-weather year, optimally timed irrigations, and a range of adjustment options, a hypothetical price increase from zero to $30 per acre-foot is estimated to reduce deliveries by half but consumptive use only by about 1%. This finding implies that a water pricing policy, or other methods of encouraging reduced deliveries, while maintaining the agriculture base, is not likely to make nearly as much additional water available for alternative urban or environmental uses as might be expected from a less realistic view of the hydrologic-economie interrelationships in irrigated agriculture. Moreover, a pricing policy would have a relatively large negative effect on agricultural net income and, correspondingly, on capitalised net income in farm land values.

Another finding of the analysis is that the estimated impact of the pricing policy depends on the particular model formulation. If adjustment options are relatively limited, the price-responsiveness of delivery demand and consumptive use demand differ little. But in a more realistic model formulation which also allows for adapting in the long run the number of irrigations and switching to more capital-intensive technologies with higher application efficiencies, the predicted consumptive use demand is significantly less responsive to increasing water prices than the delivery demand.

These results have important implications for potential water pricing policies in river basins where return flows constitute a significant part of the downstream water supplies. Our study finds that in such contexts, attempts to make additional water available for nonagricultural uses by raising irrigation water prices would be relatively ineffective in the long run because it would leave net downstream river flows and/or aquifer storage virtually unchanged. Making significant amounts of additional water available must rest on measures which actually reduce consumptive use. This will be most effectively accomplished by changes to low-consumptive use crops or to nonirrigated agriculture.

Effects of Threshold Drainage Area

HAZUS (abbreviation for HAZards United States) is a GIS-based natural hazard loss estimation software package developed and freely distributed by the United States Federal Emergency Management Agency (FEMA). The current version of HAZUS, HAZUS-MH (HAZUS-MultiHazard), can model three types of hazards: flooding, hurricanes, and earthquakes. Based on inventory data of buildings, vehicles, transportation systems, utility systems, critical facilities, hazardous materials, agricultural commodities, and demographics, as well as specified hazard parameters, HAZUS locates hazard areas and estimates the potential physical, social, and economic costs based on engineering knowledge of the damage and hazard loss estimates (FEMA

2005). Hazard loss estimation provides a basis for hazard preparation, mitigation strategy development, and response and recovery planning, and it is essential to decision making at all levels of government. Equipped with information about the potential loss, hazard managers are able to better priorities mitigation measures to reduce further disaster losses, and local communities can allocate limited resources more efficiently to responding to, and recovering from, disaster events.

This study intends to study how the HAZUS parameter of threshold drainage area affects flood modelling results. A threshold drainage area (thereafter drainage threshold) is a parameter that a user specifies to delineate a synthetic stream network from a Digital Elevation Model (DEM). It is the minimum accumulation area draining into a stream network. Decreasing drainage threshold will generate a more detailed stream network with additional tributaries. Theoretically, the minimum drainage threshold is only limited by the resolution of available DEM data (FEMA 2009a). Past studies of hydrological modelling often regard a drainage threshold as optimal if the accordingly developed drainage network matches the real-world stream network. The realistic drainage threshold may change with local relief, soil, geology, and climate conditions. Nevertheless, stream networks developed in HAZUS are for flood modelling purposes and not necessarily the same as real-world stream networks. Apparently, appropriate drainage thresholds for HAZUS flood modelling largely depend on user's need.

A small drainage threshold delineates small stream reaches and thus a maximum coverage of potential floodplains, though the offshoot floodplains may be spurious. As most hydrological studies aim to use real drainage thresholds to delineate stream networks, literature on the effects of specified threshold parameter on hydrological modelling is limited. Moussa and Bocquillon (1996) and Stepinski and Collier (2004) reported that the total length of stream networks decreases with increasing drainage threshold. Li et al. (2008) found that, as drainage threshold increases, the outflows of basins became slower, the discharge of flood peak decreases, and the basin mean concentration time becomes longer. Liu et al. (2003) indicated that using a large threshold value will result in a longer flow time and a prolonged flow response at the end of the flow path.

Another objective of this study is to investigate the effects of study region size on flood modelling. A HAZUS study region is the area defined by a user in the beginning of analysis to allow HAZUS to prepare inventory and DEM data (for delineating stream networks). Note that users are allowed to select stream reaches of interest as a hazard scenario to conduct floodplain delineation and loss estimation. We expect that different sizes of DEM data will produce different stream networks and, consequently, different flood modelling results.

Limited studies have mentioned the effects of study region size (channel network extent) on hydrological modelling. Montgomery and Foufoula-Georgiou (1993) indicated that channel network extent determines hillslope travel distances and network link lengths; the hydrologic response functions of a catchment will therefore be affected. FEMA (2009b) stated that study region size (thereafter region size) may affect drainage areas for stream reaches at the boarders of the study region and cause discontinuous stream reaches. While a stream network delineated for a larger study region appears to be more accurate than a stream network delineated for a smaller corresponding region, FEMA (2009b) recommended a four-county limitation of region size due to the extensive processing of the associated large amount of data.

As not every HAZUS flood study clarified the used parameters of study region and drainage threshold (e.g., Burton and Cutter 2008), most studies determined the parameters based on reasonable or convenient choices. The study region was set to cover the streams of interest, and the default drainage threshold of 10 square miles in HAZUS was used. Some studies have an area of interest larger than one county and set the study region to one county for computational efficiency. Considering floodplains delineated from more detailed stream networks to be more accurate, Gall et al. (2007) used the minimum drainage threshold of 1 square mile allowed by the initial release of HAZUS-MH. "Considering the tradeoff between stream network detail and computer processing time", Ding et al. (2008) determined an optimal drainage threshold of 5 square miles for a watershed of 287 square kilometres.

In delineating floodplains for all counties in Maryland, Joyce and Scott (2005) used different drainage thresholds for individual counties "without creating an overwhelming number of reaches", which are between 4 to 6 sq mi for most counties. Parameters of drainage threshold and region size are artificial factors introduced by HAZUS users that may have significant impacts on flood modelling results. While the appropriate parameters largely depend on characteristics of the study area and analysis needs, understanding potential effects of the parameters helps users to choose optimal parameters for the study area. Identifying trends and variations of the loss estimates due to changes in the parameters is also essential to HAZUS application. This study intends to establish the methodology for performing such an investigation through a case study.

Flood Hazard Modelling

HAZUS flood modelling involves analyses of flood hazard and flood loss. Flood hazard analysis consists of three stages: stream delineation, hydrologic analysis to determine stream discharge, and hydraulic analysis to determine flood depth and extent. Stream delineation generates a flow direction and a flow accumulation grid from a DEM. Stream networks are then derived comprising grid cells into which more areas than the threshold area drain.

Figure: *Flood hazard modelling*

Hydrologic analysis of stream discharge estimates the discharges at each stream's upstream and downstream nodes for specified flood return periods. The analysis uses default regression equations in HAZUS; the input variables are calculated stream and catchment statistics; and default temperature, precipitation, soil, and land cover parameters are taken into account. Estimated discharges are adjusted by default discharge data at stream gages and by default discharge frequency curves at designated main streams.

Hydraulic analysis of flood depth and extent computes flood elevations at cross-sections of stream channels based on channel morphology and the estimated discharges (for a specified flood return period). These elevations are used to form a flood elevation grid from which a flood depth grid is computed. The flood depth grid and inventory spatial data are input in a flood loss analysis to estimate possible damage to buildings, infrastructure, and key facilities.

Because building inventory data in HAZUS are summed by census block, flooded building counts are estimated by areal-weighting the general building stock within a census block. Note that this areal-weighting approach may overestimate flooded buildings, since buildings are likely to concentrate outside potential flooding areas rather than distributed equally as the model assumes. HAZUS estimates the degree of building damage by default depth-damage curves. Flooding debris and the direct and indirect economic losses are estimated thus:

- For buildings with less than 50 percent damage, the estimates are based on occupancy type and flood depth; and
- For buildings exceeding the 50 percent damage threshold, the building is assumed to be a total loss and will be demolished.

Further, HAZUS estimates shelter needs based on estimated building populations affected by flooding. To assess the effects of HAZUS input parameters of study region size and threshold drainage area, this study focuses on three HAZUS flood modelling outputs total floodplain area, total amount of flood debris, and total economic loss. Flood-generated debris may include building finishes (e.g., dry wall, carpeting, and insulation) and building contents (e.g., furniture and appliances). In severe cases, flood may damage building structure and foundation and produce debris of structural materials (e.g., framing, walls, and exterior cladding) and foundation materials (concrete slab, concrete block, or other foundation). Flood-fighting effort and the floodwater add additional debris, such as sandbags, mud, and sediment. The HAZUS flood debris model only estimates weights of debris generated from building components, including finishes, structure, and foundation; building contents and debris loads from vegetation and sediment are not included.

The direct loss estimates in HAZUS include the costs of building repair and replacement. The indirect economic loss includes the consequential loss from the immediate impact of building damage on a community, such as financial resources needed to recover from damages, financial consequences to the community's businesses due to businesses interruption, and financial costs due to potential job and housing losses (FEMA 2009b). In this study we use HAZUS to model floodplains based on a 100 yr flood return period and compare them with the 100 yr floodplains delineated on FEMA's Flood Insurance Rate Maps (FIRMs). Hundred year floodplains are areas "subject to inundation by a flood that has a 1percent or greater chance of being equalled or exceeded during any given year" (FEMA 2008a). The 100 yr floodplain is a regulatory standard used by federal agencies to administer floodplain management programs; it is also the basis for insurance requirements in the National Flood Insurance Program (NFIP). According to FEMA (2008a), the delineation of floodplains on FIRMs "is based on historic, meteorologic, hydrologic, and hydraulic data, as well as open-space conditions, flood control works, and development." In addition to FIRMs, FEMA also uses Digital Q3 Flood Data. This layer was created by scanning paper copies of effective FIRMs and digitising the floodplain boundaries (FEMA 2008b). It is worth noting that FIRMs may be outdated, particularly if local topography has changed.

HAZUS flood modelling allows for three levels of analysis. Level 1 analysis, which is based on HAZUS default data and settings, allows users to easily implement flood modelling without having extensive engineering knowledge or detailed local data, such as LIDAR derived DEM, field-measured flood-depths, and updated building inventory. However, flood distribution and loss from Level 1 analysis are relatively crude estimates and can have large margins of error (FEMA 2009b), meaning that the estimates may not be suitable for hazard mitigation decision making by local governments. According to FEMA, the estimates "should only be used as initial estimates to determine where more detailed data collection and analyses are warranted, or for studies covering large geographic areas." "This type of output would be appropriate for regional, state or national level analyses" to "provide a benchmark against which progress toward reducing losses due to natural hazards through public policy can be assessed". Level 2 analysis incorporates user-collected local building inventory, elevation, or flood depth data to improve hazard analysis and loss estimate. Level 3 analysis incorporates data and information of a specific hazard event acquired from site-specific engineering and economic studies; the HAZUS default stream discharge functions and depth-damage curves may be modified by technical experts.

This study conducts a Level 1 analysis, which allows us to efficiently implement HAZUS flood modelling for the purpose of assessing the effects of input parameters of study region size and threshold drainage area. HAZUS default elevation data, the 30 m DEM data of the National Elevation Dataset (NED), are used for developing synthetic stream networks. The National Elevation Dataset covers the entire U.S. and is accessible from the seamless U.S. Geological Survey (USGS) web site through HAZUS interface.

Studies evaluating HAZUS flooding modelling accuracy have reached miscellaneous conclusions. Ding et al. (2008) reported underestimated total floodplain areas of approximately 32 percent in a Level 1 analysis. Gall et al. (2007) argued that HAZUS is prone to under-predicting floodplains in upland or undulating terrain because floodplains not connected to a stream, e.g., a lake or pond feature, are not modelled. Contrary to conclusion of Ding et al. (2008) and Gall et al. (2007), URS Group, Inc. (2007) found overestimated floodplain areas in a Level 1 analysis; they claimed the likely reasons for this being the inability to interpret narrow channels from coarse DEMs or ponded flood water due to debris in stream channels. URS Group, Inc. (2007) also reported a 30 percent overestimation of residential loss ostensibly resulting from the even-distribution assumption and area-weighting loss estimation in HAZUS. Supporting the same argument, Ding et al. (2008) found over-estimated flood loss in a Level 1 analysis. Contrary to findings of Ding et al. (2008) and URS (2007), Meyer (2004) reported a six percent underestimation of the total economic loss from building damage in a Level 1 analysis. The different conclusions of HAZUS accuracy studies are likely due to the varied local topography and facility/housing distribution in corresponding study areas.

Our study area is the city of Ozark in Christian county, Missouri, USA, approximately 5 km wide and 10 km long, covering an area of 19.6 square kilometres. Ozark is approximately one kilometre south of the city of Springfield, the major urban centre in southwest Missouri, and approximately 33 km north of the city of Branson, a popular destination for tourists from Missouri and the surrounding states. Highway 65 is the main access mute to Springfield and Branson. As an outlying community within the Springfield Metropolitan Statistical Area, Ozark has experienced rapid growth in recent years. The population of Ozark grew from 4243 in 1990 to 9665 in 2000. In 2007, the estimated population was 17,177. This growth highlights the importance of a flood hazard mitigation plan for the city, which necessitates risk assessment and vulnerability analysis for floodplains. For Ozark, significant flood-prone areas exist in the riparian areas of the Finley River and its tributaries.

Ozark is situated in the Ozark Plateau physiographic region, which is known for a well developed karst topography. A karst landscape has land forms of shafts, tunnels, caves, and sinkholes created by groundwater dissolving sedimentary rock such as limestone. A karst landscape is scenic yet fragile and vulnerable to erosion and pollution. Frequent flooding on Ozark's karst landscape creates compound damage to the natural environment and man-made structures.

To study the effects of drainage threshold on HAZUS flood analysis for the Ozark city, we created a HAZUS study region based on all census blocks within Ozark. The USGS DEM data for the study region were used to develop synthetic stream networks. HAZUS automatically determines the needed DEM boundaries for the study region and extracts the DEM data by first selecting the watersheds (2) that intersect a 2000 m buffer of the study region and then clipping the DEM data to the selected watersheds with a 1,000 m buffer. This approach ensures sufficient DEM data for subsequent flood analysis in HAZUS. We then specified different drainage thresholds for HAZUS to develop synthetic stream networks. A total of 24 drainage thresholds from 0.25 to 6.00 square miles were tested for every 0.25 square mile increase. We then selected stream reaches that geometrically intersect the Ozark city to create the flood scenario.

HAZUS hydrologic and hydraulic analyses were conducted to generate 100 yr floodplain boundaries and associated flood depth grids. HAZUS loss estimates were conducted based on derived flood depth grids and HAZUS default inventory data. We compared the total floodplain area, total flood debris, and total economic loss for different drainage thresholds. The modelled flood plains were also compared with floodplains delineated in FEMA's Digital Q3 Flood Data. Because the selected stream reaches for flood modelling extend outside the boundaries of the city of Ozark, the modelled floodplains from HAZUS cover areas outside the city.

To compare the floodplain areas of Ozark at different drainage thresholds, we clipped the modelled floodplains to the Ozark city boundaries and calculated the clipped floodplain areas. We compared HAZUS-modelled flood debris and economic loss without clipping or other processing, understanding that HAZUS generates debris and loss estimates based solely on inventory data within the defined study region and does not consider inventory data outside the study region (FEMA 2009b).

To study the effects of region size on HAZUS flood analysis for the Ozark city, we created different sizes of study regions enclosing the geographic areas of the city. A total of 20 study regions from a size slightly larger than the Ozark city to a size of two counties (including the local and a neighbouring county) were determined in such a way that the DEM data covered an incremental number of watersheds from 2 to 21. A drainage threshold of two square miles was used to develop stream networks for different region sizes. This threshold size was selected because floodplains modelled from a drainage threshold of two square miles have a higher degree of overlap with FEMA Q3 floodplains than floodplains modelled from other drainage thresholds.

In contrast to the notably different stream networks at different drainage thresholds, the stream networks at different region sizes look identical, with variations that cannot be observed at the current mapping scale. We selected stream reaches intersecting Ozark to create the flood scenario. HAZUS hydrologic and hydraulic analyses as well as loss estimates were then conducted. The total floodplain area of Ozark was compared (for different region sizes) after the modelled floodplain was clipped to the city boundary. The estimated flood debris and economic loss were directly compared since the estimates are based on comparable stream reaches. We used the modelling results for a one-county study region as a benchmark, understanding that a large portion of hazard mitigation plans using HAZUS are prepared for local, county-level jurisdictions (FEMA 2005).

The floodplains modelled at different drainage thresholds are within a similar range of flood depth from 0.0001 to 6.74 m, while the total floodplain area (within Ozark city boundaries) has a decreasing trend with an increasing drainage threshold. The change rate is substantial, ranging from -20 to 51 percent when using the floodplain area at a drainage threshold of two square miles as a benchmark. However, at drainage thresholds of one square mile and larger, the variation decreases to within 20 percent. The decreasing trend is mainly due to the fast reduction of small tributaries when drainage threshold increases.

The decreasing trend of stream length with increasing drainage threshold is consistent with findings of Stepinski and Collier (2004) and Moussa and Bocquillon (1996). Notice that the trend approximates a logarithmic function and become relatively level at a drainage threshold of two square miles, indicating that most reaches intersecting Ozark are within a drainage threshold of two square miles. The implication is that stream reach delineation for the city of Ozark will be sensitive to a specified drainage threshold within two square miles, while stream reaches delineated based on a drainage threshold of two square miles and larger will be similar.

Compared with FEMA Q3 floodplains, HAZUS underestimated the total floodplain area at all drainage thresholds; the largest floodplain coverage modelled at a drainage threshold of 0.95 square miles underestimated 2.71 percent of the total floodplain area. The under-estimation of floodplain areas by HAZUS is consistent with findings reported by Ding et al. (2008) and Gall et al. (2007). The degree of overlapping between HAZUS-modelled floodplains and FEMA Q3 floodplains was determined using producer's and user's summary accuracy statistics.

The producer's accuracy is calculated by dividing the total overlapping area with the total area of FEMA Q3 floodplains, and the user's accuracy is calculated by dividing the total overlapping area with the total area of HAZUS-modelled floodplains. They are viewed from different perspectives: the producer's accuracy indicates the percentage of real floodplain areas which are correctly modelled, whereas the user's accuracy indicates the percentage of modelled floodplain areas that are correct.

We calculated the accuracy statistics for modelled floodplains within the boundaries of Ozark city and in its downtown areas where the concentrations of population and critical facilities are high. Accuracy assessment shows that floodplains modelled at a drainage threshold of 0.25 square miles have the highest producer's but the lowest user's accuracy for floodplains within the city limits of Ozark and its downtown area. This is because floodplains modelled at 0.25 drainage threshold cover most of the FEMA Q3 floodplains while many modelled tributary floodplains are spurious, particularly in south Ozark downtown areas. Floodplains modelled at a drainage threshold of two square miles had the second highest producer's accuracy as well as considerably high user's accuracy. Therefore, a drainage threshold of two square miles was regarded as optimal for floodplain modelling in the Ozark study area.

In contrast to the decreasing trends of the total floodplain area with increasing drainage threshold, the total amount of flood debris had an overall increasing trend, with a substantial change rate from -46 to 14 percent to the benchmark at the optimal drainage threshold of two square miles. The increasing trend of flood debris can be explained by a slower basin outflow and a longer flow time at a larger drainage threshold.

The total economic loss decreased with increasing drainage threshold, with a change rate from -20 to 24 percent to the benchmark at the optimal drainage threshold. The decreasing trend is mainly due to fewer floodplain areas as floodplains at small drainage thresholds not only cover the floodplains at large drainage thresholds but also include additional floodplains at small tributaries. It is worth noting that the impact of indirect economic loss on total economic loss becomes more important as tributary floodplains increase with increasing drainage threshold, the reason being that properties on tributary floodplains have higher economic value than do properties in the floodplains of main streams because of higher population density and business activity. The total floodplain area does not have a clear increasing or decreasing trend with changing region size, though the total floodplain area tends to be larger at a larger region size.

The total floodplain has a change rate ranging from -23.6 to 0 percent compared with that of the benchmark at a one-county study region. For a region size from 0.4 to 2.2 times a one-county area, the variation is about 5.3 percent. The total reach length fluctuates to within 7.6 percent compared with that of the benchmark as the region size increases. The small change rate is due the small change of stream networks at different region sizes. Unlike the correspondence found between reach length and floodplain area when drainage threshold changes, no corresponding changes of reach length and floodplain area were found with varying region size.

Compared to the benchmark, the total amount of flood debris fluctuated from -42 to 30 percent with changing region size. Similarly, the total economic loss did not have a consistent trend, fluctuating between -37 to 9 percent compared to the benchmark. For a region size ranging from 0.4 to 2.2 times a one-county area, the total economic loss had a small variation within 13 percent compared to the benchmark. This small variation corresponded to the small variation of floodplain area at large region sizes. This correspondence is consistent with our previous findings in the analysis of drainage threshold effects; the total floodplain area appears to be a major determinant of the total economic loss.

This study investigates the effects of drainage threshold and region size on HAZUS flood modelling. While the derived error bounds of flood estimates are only applicable to the Ozark study area, the trends of flood estimates with changing drainage threshold are likely similar for other geographic areas. This study determined the optimal drainage threshold to be two square miles by the degree of overlapping of modelled floodplains with FEMA Q3 floodplains. It would be interesting to investigate whether the optimal drainage thresholds for other study areas are the same using the proposed approach.

As the trends of estimated floodplain area and economic loss become stable at a region size of 0.42 local-county-areas and larger, a study region comprising the local county appears to be a logical, as well as convenient, choice for future HAZUS flood modelling. Stream reaches moving across the interested areas can be subsequently selected to conduct flood analysis. The study showed that HAZUS flooding modelling can be sensitive to changing drainage threshold and region size. The optimal drainage threshold was found to be two sqaure miles, while the optimal region size was determined to be the local county. The corresponding trends found for total floodplain area and economic loss with changing drainage threshold and region size change indicate that the total floodplain area is a major determinant of the total economic loss. The results of this study allow flood researchers of the Ozark area to understand the reliability of HAZUS flood loss estimates. The findings of this study also provide insights for other study areas to determine appropriate HAZUS parameters. Importantly, this study has established the methodology to perform future investigation of the effects of drainage threshold and region size on HAZUS flood modelling.

6

The Drainage Services

This chapter addresses the economics of land drainage in agriculture-a subject that is the poor and neglected cousin of the economics of irrigation. Land drainage can be defined as those processes, natural and social, by which flows of water on the land, at both surface and sub-surface levels, are conducted to other locations, such as lake, river and aquifer. The outcomes of drainage activity are not necessarily confined within the boundary of a single farmholding and therefore, like wastewater collection and disposal, drainage services may be a *public good*. A public good can be defined as any good where the benefit derived from it by one consumer does not diminish the benefit derived by consumers in general. The standard textbook example is a country's armed forces. Since no consumer privately appropriates a public good, it is often difficult and sometimes impossible to price or sell it in a free market.

In the absence of human society, the drainage of rainwater is a purely natural phenomenon. But with respect to the drainage of irrigation water and rainfall in farming areas we are dealing with flows that may require human intervention, so that we can legitimately speak of the demand for and supply of drainage services. The same holds true for urban drainage. The water inputs that source drainage water in farming are rainfall itself as well as the irrigation system's storage points, distribution network and irrigated fields. Drainage water's constituent flows are leaching applications, escapes, seepage, deep percolation and surface run-off. Drainage water's outputs are non-beneficial evapouration and transpiration, returns to the catchment's hydrological resource and the pumped flows that make up the fifth category of water appropriated for irrigation.

The demand for drainage services by farming families and agribusinesses is found whenever gravity-driven 'natural drainage', without any human operational or infrastructural intervention, does not meet farmers' needs effectively in comparison with drainage services that have been engineered. Farmers' need for drainage services takes a number of forms.

First, rainfall and irrigation water flows can collect on the land's surface at various points in the farming area. This may be due, for example, to unevenness in field level or to soil types such as clay that inhibit surface water from percolating downwards. As a result, the ground becomes waterlogged and the excess of soil water at the plants' rootzone constrains their growth and reduces tonnage efficiency. Waterlogging also occurs where the water table is shallow. Jacobs *et al.* (1997) refer to water tables as close as 50 cm to the land surface in Haryana, India, where seepage of hundreds of millimetres per year from distributaries has caused extensive waterlogging.

Second, drainage water can be reused for irrigation if that water is channelled into ditches or pipes and pumped up to field level. The additional irrigation flow is used to increase crop turnover. Here we have the fifth category of the irrigation cycle's base supply.

Third, drainage water is used for artificial recharge of the local aquifer. The additional stored volumes maintain crop output, by means of groundwater pumping, whenever rainfall is insufficient.

Fourth, irrigation water invariably contains dissolved salts. This may already be the case at the point of appropriation of the base flow, it may be introduced by farmers' application of agrochemicals, or salts may be dissolved as the surface flow of irrigation water is distributed to the plants' rootzone. The evapouration of irrigation water at the field level can increase salinity up to and beyond the concentration where it begins to inhibit plant growth. Once again, tonnage efficiency falls. Salinization also occurs where the water table is shallow, 2 m or less below the soil surface. The upward capillary movement of groundwater and its evapouration leads to salt accumulation in the soil profile. In the cases of both irrigation water and groundwater evapouration, the processes of natural drainage lower tonnage efficiency. Here planned drainage can be beneficial, including the case where water is supplied to fields and then drained precisely in order to leach salts from the soil profile. Farmers are likely to practise leaching at times of the year when the opportunity cost to them of the water thus used is lowest. To sum up, the demand for drainage services in agriculture is fourfold: to prevent waterlogging, to source one of the irrigation cycle's five base flows, to facilitate aquifer recharge, and to combat soil salinization. The first and fourth of these raise tonnage efficiency whereas the second and third augment the availability of irrigation water. In all four cases, turnover per holding is raised, provided the increase in quantity supplied does not face inelasticity in demand. Van Hofwegen and Svendsen (2000:27-8) report that about one-third of global cropland-500 million ha-is not adequately drained by natural processes. In contrast, in Asia, Europe and North America engineered drainage has contributed to raising food production on about 150 million ha. But such drainage can destroy natural wetlands and, as a result of the growth in power of the environmental conservation movements, new reclamation has slowed considerably. But much existing agricultural land still suffers from inadequate drainage in both rain-fed and irrigated areas-South-east Asia in the monsoon seasons is an example. 'In arid and semi-arid areas some 20 to 30 million hectares of land is suffering from water logging and salinity due to irrigation or high canal percolation rates'.

It is worth noting the important distinction between surface and sub-surface drainage. Their technology, infrastructural costs and effects can be strikingly different. In Finland, e this happens, both are likely to be financed by a single payment mechanism.

Just as with irrigation, we can speculate on the elasticity of demand by farmers for drainage services. Here the analysis holds. *P* would be the price of each cubic metre drained and *Q* would be the volume of water drained. But drainage services are likely to be seen by farmers more as an agronomic necessity rather than as an economic commodity. Nevertheless, Carruthers and Clark (1981:91) cite a counter-example. They state that (at the time of writing) most cultivators in Pakistan were share-tenants, 'who are, moreover, moved around by the landlords to different plots each year'. Share-tenants have a short time-horizon with respect to investment in farming; in this specific case they showed little interest in field levelling, an important technique for improving drainage and leaching.

The Supply of Drainage Services

At the heart of drainage engineering we find a means whereby the natural, gravitydriven flows of rainfall and surplus irrigation water across and beneath farmland are substituted by man-made channels powered by gravity or pumps. As a result, from a civil engineering and hydraulics perspective, the supply infrastructure of drainage services bears a striking similarity to the abstraction and distributional infrastructures of irrigation supply.

The channels are dug ditches without lining or lined channels or pipes. In Chapter 3 we have already noted the efficiency of axial flow pumps in lifting large volumes of water at low pressure through a few metres, a common case in land drainage. Deep tube wells may be employed, as in Pakistan's Punjab province, to pump water for vertical (rather than horizontal) drainage in order to reuse the flows, perhaps after adding river water to reduce the salt content of the conjunctive supply. Vertical drainage may also take place to lower the water table and drive salts in the soil below the rootzone of the plants, where they can be harmlessly stored (Seckler 1996:15). The common features of the civil engineering of land drainage and that of irrigation water abstraction and distribution have important consequences in economic terms. Land drainage networks may require major capital expenditure and capital financing. In the short term there are substantial economies of capacity utilisation. In the long term, from the planning approach discussed in section 3.3, we can also expect economies of scale because of the indivisibility of drainage infrastructure provision. Lumpiness is also likely to be a feature of drainage networks. The double punch of indivisibility and lumpiness means that mediumterm and long-term planning for drainage offers considerable benefits.

Pumping stations need to be operated and their equipment maintained. Drains require continuous work to remove debris of all kinds, to dredge silt and to cut back plant growth in and on the edge of watercourses. The mowing of bankside vegetation provides control of weed growth and encourages a healthy root system, stabilising the banks and giving some protection against erosion and slippage. Aquatic vegetation assists oxygenation of drainage water and channel stabilisation and also provides habitat for fish and invertebrates. But in excess amounts such vegetation holds water back, raises the water level, encourages siltation and leads to flooding. In Norfolk channels are desilted only when 300 mm of silt has accumulated. A channel's centre section is desilted over its entire length but the two margins are left untouched. 'This method is preferable from the land drainage standpoint as it allows continuous unimpeded central flow and will also assist future conservation by the creation of pools and shallows at the water's edge'. Hydraulic excavators, draglines and tractor-mounted mowers may be used to automate the maintenance process. Sluice gates require sluice-keepers or may be operated remotely using telemetry systems for monitoring purposes.

Currently a strong interest is shown in controlled drainage strategies. These are practices that allow the farmer to trigger drainage only after the groundwater level in a field has risen so far as to threaten crop yields or prevent salt leaching. The associated reduction in drainage flow under such a regime permits a lower rate of abstraction for irrigation purposes. To the degree that unnecessary overdrainage is reduced, more water becomes available for downstream irrigators or for non-agricultural abstractors. The two main methods for drainage control are:

1. The blocking/unblocking of lateral or collector pipes using a purpose-designed device. Here the water table rises and falls in response to irrigation releases and switching drains off or on. Farmers in some areas have used mud and straw as the control device but this can lead to system blockages.

2. The placing of a (fixed or adjustable) weir in the drainage ditch or sub-surface drainage system. Here the drainage flow takes place only when rainfall or irrigation inputs bring the water level above the top of the weir.

It is suggested that controlled drainage can be beneficial in many arid and semi-arid regions of the world where water tables are high, such as Egypt, India and Pakistan. In the Nile Delta controlled drainage trials under rice have achieved large water savings and farmers are being encouraged to adopt the technique. In 2000 Catherine Abbott and her colleagues at HR Wallingford and the Drainage Research Institute, Cairo, were carrying out field trials on controlled drainage of dry-foot crops to assess potential water-saving benefits. The work will culminate in 2001 with the development of predictive tools to help assist in evaluating the costs and benefits of this approach.

Drainage Water Quality

Modern agriculture makes extensive use of biocides and fertilizers. The former protect cultivated plants from a great variety of natural pests, the latter promote plant growth, particularly by means of increasing the supply of the macronutrients nitrogen, phosphorus and potassium at the rootzone. As a result, the farmland drainage water that returns to the hydrological resource is more or less polluted. The introduction of synthetic chemicals began in the 1940s-at least in the UK. Since that time many, many hundreds of products have been developed for use, particularly in the case of fungicides, herbicides and insecticides.

Figure: *Water Quality.*

Their use varies considerably from region to region, and in the more developed farming areas such as the UK it is not uncommon for six or seven different chemicals to be applied in a single season. In most countries our understanding of the scale and type of pollution is deficient because of the variety of chemicals and the costs of water quality monitoring. For the same reason, the impacts on human health and ecosystems are not well understood. This is exacerbated by the likely difference between short-term and long-term impacts. As Falconer (1997:16) points out: 'An important distinction must be made between point sources of contamination (e.g. from yard or storage spillages, and careless or illegal disposal) and non-point sources, such as the diffuse emissions resulting from normal, legal field use ...chronic and diffuse problems are generally not reported due to their invisibility.' The policy measures that have been adopted to mitigate or eliminate the agrochemical pollution problem are varied. They include:

1. the promotion of organic farming-Keveral Farm is an example;
2. bans on the manufacture and use of specified products;
3. the production of low-dose and less persistent chemicals;
4. the levy of pesticide taxes as well as payments for the non-use of agrochemicals;
5. information and training for farmers on the safe use, storage and disposal of products;
6. specification of exclusion zones for agrochemical use around aquifer recharge areas, along watercourses and on field margins;
7. the setting of minimum water quality standards such as for drinking water;
8. the introduction of more comprehensive and more effective freshwater treatment.

As we have seen, drainage water is an important medium for the diffusion of agricultural pollutants beyond the farming sector and its particular concern for soil salinization. The policy issues concern either the agricultural use of agrochemicals or the standards and technologies for the treatment of abstracted water destined for households and industry. Thus the subject area, however vital, essentially falls under the economics of agriculture and the economics of freshwater treatment, outside the field of the economics of irrigation and drainage. For that reason, in this book fertilizer and biocide contamination of water by farming families and agribusinesses will be addressed only in so far as it influences the efficiency of irrigation water use.

Case Study: Farmland Drainage in Norfolk, UK

The county of Norfolk is located in the east of England and the waters of the North Sea wash its coast. Norfolk is renowned for its low topography, the fertile soils of its farmland and the rich diversity of its wildlife habitats. Annual rainfall is plentiful so that, with the flatness of the terrain, the inhabitants of the county for thousands of years have had to drain the land in various ways. This is of special importance to farming families and agribusinesses in order to prevent waterlogging and flooding. Important crops are wheat, barley, rye, oilseed rape, linseed, potatoes, parsnips, sugar beet, other vegetables and salad crops. Cattle grazing is widespread.

The advantages of co-operation in providing drainage services have been recognised in the area since at least the time of the Roman occupation of Britain. The modern history of drainage dates from the 1930 Land Drainage Act that required internal drainage boards (IDBs) to be established in low-lying areas of England and Wales to provide a common service. In each designated drainage district a board was formed from members elected by local ratepayers and local authorities. In the case of Norfolk a step of major importance was taken in 1967 with the creation of the King's Lynn Consortium of Internal Drainage Boards ('the Consortium'). Initially this brought together eight IDBs.

An important development took place in 1991-4 with successive Land Drainage Acts imposing environmental and conservation duties upon IDBs and providing for codes of practice on these matters. By the year 2000 the Consortium acted for sixteen IDBs. Its area in 2000 was 160,000 ha, it had forty-three pumping stations and it was responsible for 3,500 km of watercourses. The IDBs' locations are closely related to the position of the 20-foot (6.1 m) contour in Norfolk. The core duties of an internal drainage board are to secure the efficient working and maintenance of existing drainage works in its district and the construction of new works as may be necessary.

There are powerful economic and political reasons for the IDBs to work together within a consortium. Short-term and long-term economies exist in infrastructural provision. For example, it becomes financially feasible to purchase machinery such as a £100,000 hydraulic excavator used for drain construction ('drain' is the generic term used by the Consortium for its drainage channels). The equipment is fully used and its size avoids the costs of indivisibility. Similarly short- and long-term economies are to be found on the OM&M side, as well as with planning, finance, information technology and civil engineering skills. A single drainage district, for example, could hardly contemplate employing on a full-time basis its own graphical information system (GIS) expert or its own environmental scientist. The Consortium's total employed staff in 2000 was fifty-three-made up of eight professionals, six administrative staff and thirty-nine employees in the civil works direct labour force.

The presence of a group of full-time professionals in the Consortium's headquarters with an annual expenditure of about £5 million and responsibility for 160,000 ha also gives the group political clout with bodies such as the Environment Agency, the Ministry of Agriculture, Fisheries, and Food (MAFF), the Department of the Environment, Transport and the Regions, the Broads Authority, English Nature and the Royal Society for the Protection of Birds.

The specific focus of this case study is the West of Ouse Internal Drainage Board, located in the fenland area of west Norfolk between the towns of Wisbech and King's Lynn. West of Ouse is as flat as a pancake, consists entirely of farmland and about twenty villages, measures some 13,370 ha and is made up of land largely reclaimed from the North Sea between AD50 and 1978. The Environment Agency's long-term average rainfall data for the period 1961-90 varied between 553 mm and 595 mm across the seven points where it is measured. The modal value is 579 mm. Over the same period Meteorological Office data show that effective rainfall averaged approximately 149 mm. The scale of irrigation in the area is not great and so drainage is primarily of rainfall run-off in order to prevent waterlogging. The water table is close to the land surface in this area. There is no reuse of drainage water for irrigation purposes. In general in Norfolk, irrigation takes place most on the higher, silted lands.

The drainage infrastructure is 620 km in length, almost entirely composed of excavated, unlined earth drains. One-third of the area, in the north-east, drains by gravity to five outfalls located on the River Great Ouse. The other two-thirds of the area, in the south-west, drains by gravity to Islington Pumping Station whence the drainage water is pumped up to One-Mile Drain from which it flows by gravity into the Great Ouse, which is still tidal at this point. One-Mile Drain also acts as a storage reservoir prior to discharge at low tide. All the outfalls to the river have pointing doors and are powered simply by the relative movement of drainage water and the river's tides. Islington Pumping Station has three diesel engines driving axial flow pumps and, since 1993, two submersible electric motors driving two mixed-flow pumps.

The Consortium staff maintains for the West of Ouse IDB an income and expenditure account as well as a balance sheet. This includes outgoings in respect of both farmland and village drainage. Just four items make up 87 per cent of the total. *First,* the Environment Agency precept of £176,000 is the payment due from West of Ouse to the Agency for the services it provides to this IDE. The most important of these services are the Agency's defence of the Wash coastline and the control of flooding from the River Great Ouse, primarily by means of embankments. *Second,* we have £121,000 in loan charge payments on the finance raised for various infrastructural works such as the Smeeth Lode channel and pumping plant. The UK's Public Works Loan Board (PWLB) sources such loans. *Third,* OM&M costs amount to £238,000; the item 'Use of Consortium plant and labour' is West of Ouse's largest single entry. *Fourth,* the miscellaneous costs of administration total £114,000, the bulk of which is incurred at the Consortium's headquarters in King's Lynn.

The Consortium's expenditure on capital goods is funded by MAFF grants, Environment Agency grants, PWLB loans, internal funds and leasing arrangements with machinery and equipment suppliers. Leasing rather than outright purchase may secure Agency subsidies not otherwise available. The final issue to be discussed in this case study is the way in which the Consortium's current expenditure on farmland drainage is financed. I believe the principles could be of great interest internationally. To begin with we should note that, whereas the Consortium's activities embrace both urban and farmland drainage, outlays in villages and towns are met from charges paid by local councils. Farmers are charged only for farmland drainage. It is also the case that in any single year, such as 1999-2000, farmland drainage costs are pooled across all the sixteen IDB members of the Consortium. The current costs of each single IDE, such as West of Ouse, can be identified, but that individual board's outlays are *not* the basis of that board's farmers' payments. Instead, *all* the Consortium's current expenditures on farm drainage are aggregated and this sum is then divided for payment between all the farmers located in the Consortium's area.

But what is the principle on which this division of pooled costs between farmers is made? At this point we need an historical perspective. In Britain the taxation of real property to generate local revenue is a practice of considerable antiquity and their formal exclusion by the Poor Rate Exemption Act of 1840.

Table: *West of Ouse Internal Drainage Board projected expenditure on current account 1999-2000*

Item	**£**	**%**
Environment Agency precept	175,894	24
Bridge surveys and repairs	10,000	1
Loan charges:		
Smeeth Lode channel works	16,182	
Smeeth Lode pumping plant	44,499	
Reeds drain	10,406	
Walpole west drain	35,130	
Shire drain		14,377
Subtotal	120,594	16
Telemetry works	10,850	1
Strategy investigations	6,500	1
Operation and maintenance costs:		
Use of Consortium plant and labour	210,000	
Materials	16,000	
Hire plant, compensation, etc.	10,000	
Maps, safety equipment, etc.	1,700	
Subtotal	237,700	32

Pumping costs:		
Fuel, repairs, etc.	6,500	
Electricity	16,000	
Operators	5,000	
Insurance	5,000	
Subtotal	32,500	5
Miscellaneous engineering costs	33,740	5
Property and administration:		
Insurance, water rates, etc.	1,039	
Repairs and materials	3,000	
Consortium charge	98,500	
Travelling, members' expenses, subscriptions	1,850	
Indemnity insurance	1,669	
Printing, stationery, legal fees	750	
Contingency allowance	5,000	
Audit fee	2,000	
Subtotal	113,808	15
Grand total	741,586	100

Source: reproduced with permission from KLCIDB, Accounts Department.

A tax on immovable property is particularly suitable to the functions of the local state since evasion is extremely difficult and there is no problem in attributing the yield to even the smallest units of local government, as in the case of internal drainage boards (Merrett 1986). The tax can be levied in the household sector, on industry and in agriculture. Only the last case is considered here, although the taxation principle was originally the same for all three sectors.

Within the relevant area (in this case the hectarage of the sixteen IDB members of the Consortium) each agricultural holding is separately recorded on a valuation list. In England and Wales the Local Government Act of 1948 gave the responsibility for valuation to the Inland Revenue, where local valuation officers carry it out. These officers then calculate for each holding the annual sum for which the land and buildings on it could be rented on the open market. It is assumed that the costs of insurance, maintenance and repair of the property are paid for not by the owner but by the tenant. This estimated annual rent is called the property's *net rateable value* and is set at a defined number of pounds sterling.

Specifically with respect to the net rateable value (NRV) of agricultural holdings-both the land and the farm buildings-these are based on what the Agriculture Acts call a 'full agricultural tenancy'. They were last assessed in

1991 and were carried out on contract for the Inland Revenue, mostly by private valuers but also by some public-sector district valuers. There has been no general reassessment since that time but provisions exist in the 1991 Land Drainage Act for a farming family to seek revaluation because of a change in the circumstances of a holding, for example the removal of greenhouses. The cost of the valuation process, a transaction cost, should not be neglected by any government considering introducing such arrangements.

Rates due for farmland drainage are set as a defined number of pence. For every 1 pound sterling in the NRV of a property, this number of pence is payable. (Since there are 100 pence in the pound, rates in pence can be equally well thought of as a percentage of the farm's NRV.

That was not true before decimalisation took place in the UK in 1971, at which time there were 240 pence in the pound.) In a simple case where the rate is 5 pence, and Farmer Giles has a small property with an NRV of £1,000, the rates he would pay are £50, i.e. 5 pence in the pound multiplied by 1,000 pounds. As befits a wealth tax, the more valuable the holding in rental terms, the higher the rates and the greater the contribution made to drainage costs. The payment of rates is made by the occupier of the farmholding, for in its developed form it is regarded in law as a tax on the benefit of occupation of land and buildings.

In 1999-2000 the rate charged for land drainage in the Consortium's area was 6.77 pence. Because the NRV of farmland and farm buildings at the beginning of that year for the West of Ouse was £8.9 million, the total sum raised by the rates was £603,000. The specific costs on current account of farmland drainage in that year were £742,000. So the income from rates covered 81 per cent of drainage costs. The collection effectiveness of the tax is phenomenal; in the Consortium area 99.9 per cent of the tax owed was actually received.

Because drainage costs are pooled across the sixteen members of the Consortium, the drainage rate paid in one IDB, such as West of Ouse, may be equal to, higher than or lower than its actual drainage costs in any one year. The institutional skills that hold the Consortium together as a co-operative body show an understanding that, in the cycle of capital works and OM&M over the years, no one area will be singled out for special favour or disadvantage. The Consortium's Board of Management represents all the individual IDB s and the golden rule-'everyone gets his fair share', equal gain in the long term-is watchfully, if not systematically, monitored.

Drainage Needs in the Lower Indus, Pakistan

The Left Bank Outfall Drain (LBOD, stage I) project in Pakistan provides a case study of a major investment to provide drainage services in an irrigated command area. It illustrates many of the physical and economic characteristics of drainage design and implementation and of the application of cost-benefit analysis to such investments.

The problems of rising water tables resulting in waterlogging and salinization of soil are nothing new. About 4,000 years ago, the Indus river civilization that had flourished for 1,000 years at Mohenjodaro in what is now Sindh province came to an abrupt end. Some archaeologists have speculated that prolonged irrigation without drainage had caused the water table to rise. Floods and soil salinity then destroyed the agriculture on which the city depended. Other ancient civilizations, apart from those in the Indus valley, were also afflicted, for example those that grew up on the flood plains of the River Tigris and River Euphrates in what is now Iraq.

In Sindh a major programme of improvement and construction of inundation canals was undertaken in the latter half of the nineteenth century, but it was not until 1932 that barrage-commanded irrigation was introduced with the construction of the Sukkur barrage system. This commanded a gross area of some 5 million ha on the left and right banks of the lower Indus. Later two other barrages, Kotri (1955) and Gudu (1962), completed the system as it is today (WAPDA 1997). The LBOD stage I project addresses the drainage needs of the left bank of the Sukkur barrage command. This is almost entirely a perennially irrigated area supporting a kharif crop of mainly cotton, and a rabi crop of mainly wheat. In 1932 it was known that drainage would eventually be required, but the deep water tables at that time meant that it was not initially needed. Thus the LBOD project might be regarded as a deferred part of the original investment in the irrigation system itself that commenced with construction of the Sukkur barrage in 1932!

The water table in irrigated areas in Sindh has risen by as much as 27 m in the past 30 years and over 3 million ha are waterlogged. Although rainfall averages only 125-150 mm/year, 50mm may fall within 48 hours in summer storms. Gradients on the floodplain are low and after storms water may stand in the fields for weeks, creating a public health hazard and damaging crops. Salinity, which affects over 25 per cent of the farmland, occurs when the high water table brings salts into the evapouration zone of the soil profile. Farmers dare not allow their land to lie fallow for fear that salt will encroach, and in parts of Sindh 'islands' of green fields are surrounded by a wasteland of abandoned land.

In this environment, continued diversion of Indus water without drainage will inevitably result in waterlogging and salinity, and this in turn will lead to depressed crop yields and further loss of cultivable land. If water is added without drainage, the water table will rise, and salt will accumulate if that water is saline. On average 1 ton of salt is added to every irrigated acre in

Sindh each year. With summer temperatures well in excess of 40C, water is drawn up through soil capillaries and evapourated within the crop rootzone, leaving behind the salt. Before 1932 all of the stage I project area had a water table less than 3 m deep; 50 years later 75 per cent of the area had a water table less than 1.5m deep, and a further 20 per cent less than 2.5 m deep.

Seasonal variations in water table levels are dependent on rainfall, whereas the degree of soil salinization depends on water table depth and irrigation applications. Water moves salts either upwards through capillaries or downwards by deep percolation. The closer the water table is to the surface the higher the evapouration, while for deep water tables capillary transfer and evapouration decrease rapidly. Water table depth tends to stabilise where a balance with evapouration is reached. With no drainage and projected future water supplies, this would have been at about 1 m deep (WAPDA 1997). Better water availability as well as seepage lead to waterlogging being concentrated along major canals, while tail-end areas usually experience water shortages and typically have deeper water tables.

The salinization process can be rapidly reversed by irrigation. Although irrigation is responsible for the long-term build-up of salinity, it is nevertheless sufficiently salt free (about 400 p.p.m.) for irrigation applications. With sufficient application, downward percolation will remove not only the delivered salt but also salt accumulated by capillary evapouration. Thus under intensive irrigation salinity can be controlled as long as drainage maintains the water table at a safe depth, or keeps the groundwater moving to prevent salt build-up. For fallow or uncultivated land with a water table near the surface, water movement is upward and salt accumulates in the soil profile. For such an area adjoining irrigated land 'dry drainage' may also occur. This is when irrigation water percolates to the water table, moves laterally and is evapourated up through the soil of the uncultivated land, thus causing rapid salinization. The two functions of drainage, water table control and removal of flood water after rainfall, are also not independent of each other because of the additional capacity of the soil to absorb stormwater when the water table is lowered.

In designing drainage the depth at which the water table is to be maintained is inevitably a compromise between what is desirable and what is essential and affordable. The deeper it is the better the drainage, the better the ability to cope with normal fluctuations, and the more storage is available to accommodate rainfall.

Data on the relationship between crop yield and water table depth are scarce, but field studies in Sindh resulted in relationships being established for the principal crops. Deeper rooting crops such as cotton need about 1.8 m of unsaturated soil for full root development. It is also susceptible to damage from flooding. As noted above, the salinization of soil through evapouration from the capillary fringe is also substantially less at greater water table depths. Negative aspects are that some crops draw on the capillary fringe under conditions of water supply shortage. Moreover, in areas such as rural Pakistan hand pumps may be the main source of domestic water supply.

The LBOD project is designed to stabilise the water table at 2 metres or more, halt salinization and remove storm flooding more quickly. With drainage assured, farmers should be able to increase irrigation, achieve higher yields, reintroduce fallow into the crop rotation and face lower risks. Despite this apparent need for drainage its economic viability must still be assessed. Economic viability depends on being able to show that a greater planted area and/or higher yield of crops can be achieved after the land has been drained and has a sufficient incremental value of crops to pay for the costs involved.

There are other benefits to the infrastructure but these are relatively small and/or difficult to quantify, for example improved road communication via drain embankments, reduced subsidence for existing roads and buildings and secondary economic benefits in the rural economy. Cost-benefit analysis for the LBOD project showed that because of its high capital and recurrent cost drainage alone would show an inadequate return on investment. Rates of return for drainage investment alone in irrigation schemes are always likely to be low because the primary benefits-losses in productivity avoided-materialize gradually over a long time period. The economic viability of the project thus hinges on increased yield from a larger area of crops gained by rehabilitation and improved management of the water distribution system and of on-farm irrigation methods.

The planned drainage system has been described as a mirror image of the irrigation system. Stage I is the first phase of the project to be implemented, covering over 0.5 million ha of the left bank command, and necessarily including construction of the large spinal drain which will take drained saline water to the sea. The infrastructure consists of the installation of drainage tubewells to extract groundwater hence lowering the water table, and a network of surface collector drains that feed into the spinal drain. These will also act as interceptor drains to speed the removal of storm floods. A small area will be drained through the installation of sub-surface tile drains because the aquifer there is unsuited to tubewells.

Horizontal drainage in the form of tile drains was found to be substantially more expensive than vertical drainage using tube wells: combined capital and recurrent costs in terms of present value per hectare over the life of the scheme were approximately ten times greater. It is also only practicable to lay tile drains from about 1.8 to 2.5 m, as beyond the latter figure they become much more expensive. Under irrigated land the water may rise to 1.5 m at the mid-point between tiles but in fallow conditions will lie at an average of 2 m.

The water to be drained in the area is mostly highly saline; effluent from tile drainage can vary from 3,000 to about 10,000 p.p.m. dissolved salts, and that from tubewells can be twice as saline (WAPDA 1997). A major design problem was to remove the water without affecting good agricultural land, but disposal is difficult-the choices being the Thar Desert (east of the project area), the River Indus, the Rann of Kutch, or the Arabian Sea. Disposal into the first two was ruled out because of cost and hydraulic and environmental considerations. Outfall into the Rann of Kutch (a low-lying area of salt marshes and tidal lagoons) or the Arabian Sea was technically feasible, but the latter has the better hydraulic performance and environmental impact. In addition it isolates the discharge from Indian Territory. This latter point illustrates the possible political difficulties of disposing of drainage water, a mirror image to the difficulties of sharing river waters that cross international boundaries.

Drainage should halt the loss of cultivable land and sustain yields, but additional irrigation supplies are necessary for significantly increased yields and expansion of crop areas. As noted these gains are critical to the economic viability of the project. Additional irrigation supplies will come from measures taken within the project area and from sources outside. In the project area one of the best methods of providing the farmer with more water is to avoid wasting it in the first place. This is the function of the on-farm water management (OFWM) component that encourages the realignment and partial lining of watercourses (final distribution channels managed by farmers) to reduce seepage losses, and also assists in precision land levelling to improve the efficiency of watering. A free design and supervision service plus a subsidy for 75 per cent of the cost of materials was made available to farmers for watercourse improvement. The objective during LBOD stage I was that half the watercourses in the project area should be improved.

Along major canal lines, scavenger tubewells capable of separately pumping fresh (non-saline) groundwater from a lens that may overlay more saline deeper water, and interceptor drains that run parallel to canals, recover some of the inevitable seepage losses for irrigation use.

Outside the project area Chotiari lakes are being used to provide local storage to regulate supplies while the Rohri and Nara main canal systems have been remodelled to carry additional supplies from Tarbela dam and (if built) the proposed Kalabagh dam. Together these inputs were predicted to result in a 50 per cent increase in water availability and a 40 per cent increase in cropped area, the difference reflecting the present degree of underwatering or water spreading. Note that where farmers are not badly affected by high water tables they tend to apply less than the recommended amount of water to crops, spreading it to cultivate a larger area. Although this reduces yields per unit area, it is the optimal strategy for profit maximisation as it maximises returns per unit of water, which is scarce compared with land. The practice does run the risk of not meeting crop needs if later deliveries fail, or not meeting leaching requirements and thus contributing to soil salinization.

The LBOD project concept described above was subject to intensive project preparation for more than 20 years. The definitive Lower Indus project study made the first proposals for a drainage project in 1966. The first full feasibility report for LBOD was produced in 1969, followed by a second project planning document in 1972. In 1975 the government of Pakistan began construction of the spinal drain using its own financial resources, but as a result of rising costs the project was revised and reduced in scope. In 1980 a study was undertaken with technical assistance provided by the UK's Overseas Development Administration that identified the LBOD component projects, established their priorities and recommended an overall implementation plan. By 1982 when the World Bank took an active interest in the project some 100 miles (160 km) of the spinal drain had been constructed including an outfall into the Rann of Kutch. In 1982 the World Bank, concerned that the project had moved from a multi-purpose drainage project to a single-minded determination to construct the spinal drain-something they suspected would not be viable on its own-engaged consultants to carry out a fully detailed project preparation study for stage I of the LBOD programme. Although this long history of planning studies and reports was not particularly planned or co-ordinated, it does mean that LBOD stage I has evolved through a series of investigations, appraisals and processes of refinement during which it has been subjected to extensive professional scrutiny. This partly reflects the scale and importance of the project (to drain over 0.5 million ha at a predicted cost of US$636 million) but it is also an example, if a lengthy one, of how project preparation and appraisal in the water sector is likely to be an iterative process. The process may be further complicated by the complex negotiations and arrangements required to secure the financing of such a large project.

For the economic and financial cost-benefit analysis of the project the major cost items were the civil engineering costs of installing the drains and tube wells, and their identification was thus straightforward. Identifying the benefits was more difficult-a characteristic problem for investments in drainage for which a large proportion of the benefits may derive from predictions of future losses in productivity that will be avoided. As described above the general contention is that the LBOD project will increase yields and cropped area over approximately 0.5 million ha of command by lowering water tables and salinity levels, saving or recovering water lost from the irrigation system, providing supplementary irrigation water and reducing flood damage. A shift toward more diversified high-value crops, such as vegetables and orchard fruits, was also predicted. Incremental benefits depend on higher production in the 'with' than in the 'without' project scenario, depending very much on changes in land use. Without the project, rising water tables caused the intensification of cropping by farmers on their best land, as fallow areas were reduced and land abandonment increased. It was projected that further increases in irrigation supply would initially increase the cropped area but at the expense of non-cropped area, reducing the dry drainage effect and allowing the water table to continue rising. This would force the farmer to intensify further and all fallow land would potentially be lost. Eventually the water table might rise to a level that inhibits cropping and cropped area would begin to reduce. Any further increases in supply would then cause wholesale abandonment of land. At the time of the project preparation study some areas had reached this point with pockets of cultivated land surviving because a saline waste surrounded them, providing the dry drainage that kept the water table just deep enough for cropping.

An initial study of marketing and pricing of agricultural products led to the conclusion that existing marketing, processing and pricing arrangements for the project's agricultural inputs and outputs were satisfactory and not likely to impose constraints on achieving projected production and farmer income levels with the project. High-yielding varieties and fertilizers had already been widely adopted, and it was expected that increased water supplies would soon be taken up both to increase applications to existing cropping and to extend the crop area. The value of surface drainage was also clearly understood by farmers, who commonly ran off stormwater to the few drains that existed or to nearby depressions. Less well known was their attitude to subsoil drainage, although there were some indications that the need was realised. Overall it was assumed that the farming community would respond well to the provisions of the project and the expected benefits would thus be generated. However, almost completely lacking from planning and the early stages of implementation was any participation by farmers in the project design or even an effective communication strategy to explain what was intended.

To date it is difficult to evaluate the success and impact of the LBOD project. Delays in implementation meant that drainage tubewells only began to operate over a significant part of the project area during the late 1990s. Operation and maintenance problems also inhibited the full and continuous operation of those works completed. Where operation has been sufficiently continuous there is localisation evidence that reversing waterlogging and salinization has improved the sustainability of irrigated agriculture, while the change from dry drainage to drainage with disposal has enabled additional irrigation water to be used effectively and the area of abandoned land to be reduced. These improvements have had a positive, though modest, impact on land values, agricultural production and rural employment. Evidence of yield improvements attributable to the project is much more limited. Many farmers still fail to invest sufficiently in the inputs needed to raise yields, in part because of the on-going failure by government agencies to deliver both irrigation and drainage services that are adequate and reliable. Other important factors have been poor price incentives, particularly for wheat during the 1990s, and poor performance in the delivery of other services such as credit, certified seed and extension. An important lesson from this case is that drainage and irrigation delivery improvements alone will not necessarily improve the performance of agriculture if farmers also face other constraints.

It is also clear that effective strategies for operation and maintenance and for cost recovery are critical. This is difficult because for drainage there are additional problems to the normal difficulties of revenue generation from low-income farmers through irrigation charges. Farmers may regard drainage as not directly productive and that it should be a part of national infrastructure that is a government responsibility like roads. Indeed they may expect compensation for land lost to surface drains rather than levying of charges. Those downstream may claim that it is salt disposal upstream that is creating their need for reclamation and drainage. Upstream farmers are likely to be reluctant to accept a share of downstream drainage costs even though, as in the case of the LBOD, they are dependent on downstream maintenance of the surface disposal network. Current drainage problems were also created by past decisions to defer drainage investment. The costs avoided have accrued as a benefit to either government or farmers who have cultivated without drainage. Should current farmers now pay the full costs of drainage or should government pay for its previous short-sighted investment decisions and deficient regulatory policies? How should such choices be weighed in the context of the other subsidies or taxes applied to other farm inputs and outputs?

Drainage can often be viewed as closely approximating the economic characteristics of a public good, i.e. non-excludability and non-rivalness. There is a joint or collective demand for drainage, and if regional or area drainage is provided (as with the LBOD) no farmers (or other residents) can be excluded from benefiting even if they pay no drainage charges. Once water tables are lowered across the area non-rivalness applies, as enjoyment of the benefits by one does not consume or diminish the benefits for others. Assuming farmers can afford the investment needed, agricultural drainage can be a private good if farmers can drain their own land and either dispose of the drainage water on their land or reuse it for irrigation. Neighbours may still receive a positive externality from the lowered water table, or a negative one if disposal causes environmental damage. However, once collective disposal systems are needed public good arguments again apply. For the LBOD project the need to get rid of saline groundwater creates the requirement for a regional infrastructure of surface drains. Even for these excludability would be difficult and costly to enforce, and non-rivalness applies as long as capacity is sufficient to accept the disposals from farmers.

In schemes such as the LBOD neither the benefits nor the infrastructure are evenly distributed, yet all components must be well maintained and operated for all to benefit. The costs of drainage cannot easily be subdivided and allocated, particularly where farms are small and the hydrological boundaries for the effects of tubewell drainage hard to determine. Similarly it is not easy to see a system of subdividing the benefits into purchasable units that can be competitively sold separately to different individuals. In other words, property rights-the basis of all markets-cannot easily be established for regional drainage. Here then is the drainage challenge. Essential for sustainable irrigation in many areas, economic theory suggests that public provision is necessary (at least for major elements of the infrastructure), but experience as exemplified by the LBOD suggests that developing countries have overextended public-sector commitments exacerbated by low efficiency and accountability of public agencies.

It is no mean challenge. For example, combined irrigation and drainage operation and maintenance costs for the LBOD scheme may be ten times greater than those for irrigation alone. In fact, farmers have been paying less than half of the irrigation costs. If pre-project predicted levels of benefits are achieved farmers should be able to pay such costs from incremental income, but as noted actual achievements may be rather less. Not adequately addressed during planning or early implementation, these issues are being tackled in the LBOD scheme through a number of pilot initiatives. Improved communication with the population of the project area and participation by farmers in the design and construction of on-farm works such as tube well disposal channels has brought improvements. Farmer groups may be able to take responsibility for some operation and maintenance tasks but modalities for this are yet to be resolved. Use of private contractors and performance contracts for operation and maintenance has been trialled successfully, aiming to minimise the number of staff employed by government agencies while ensuring that equipment and workshops already supplied under the project are effectively used. Despite the importance of the scheme these and other unresolved funding and management issues threaten its success and sustainability.

Public Goods and the Christian Eucharist

This chapter began by introducing the concept of a 'public good', i.e. any good where the benefit derived from it by one consumer does not diminish the benefit derived by consumers in general. The term deserves some discussion. Curiously, 'public good' cannot be applied to any goods at all, at least in the sense of a good as an artefact, for clearly goods in that last sense *can* be privately appropriated. It is also important to recognise that the supply of a public good does diminish the benefits derived by consumers insofar as it draws down on resources that could have been applied to alternative goods and services. The competition for resources between guns and butter (national defence versus national consumption) is the starting point of chapters on the production possibility curve in introductory textbooks on economics. Nor should we confuse a public good with a publicsector good or service-the institutional ownership of the supplier has nothing whatsoever to do with the economic definition of a public good. Drainage, for example, is a public good whether or not it is supplied by a district board or a private company. It might be thought that a public good is one where the benefits to any single consumer cannot be identified. This too is false. For example, the treatment of polluted industrial wastewater prior to its discharge into a river is certainly the provision of a public good, but the change in the quality of river water abstracted by any single actor can certainly be carried out in great detail.

I am now convinced that the theoretical economist's concept of a public good is a cure for which, unfortunately, there is no disease. What *matters* is that drainage and wastewater treatment once in supply can benefit a variety of catchment actors, the services commonly enjoyed in this way are non-exclusive, and the measurement of the uptake of the service with its consequential advantages may be complex, open to dispute or just downright costly.

So the payment system for this service supply requires a design quite different from that of the bulk of goods and services in terms of both assessment and collection. It is this that makes the English and Welsh system of NRVs of such interest with respect to drainage. Within the last year or so Swedish institutions have begun financing studies of water resource management in catchments shared by more than one nation from the point of view of such management importantly constituting a public good. The resources applied would be better used to revive the theological discussion of transubstantiation.

Institutional Economics and Irrigation Policy

This chapter has two objectives. The first is to provide a summary of the book's institutional economic content-to recapitulate the broad lines of the text's economic reasoning.

The second objective is to make proposals, not about irrigated agriculture in general (for which I am not qualified) but about the application of institutional economics to irrigation policy, including the design of policy instruments.

Chapter suggests that the economics of irrigation is the application of economic theory to the irrigation cycle. The hydrosocial activities within this cycle are the appropriation of water for irrigation, its storage, its distribution to and from storage, its use in farmers' fields and the drainage of the water not consumed in evapotranspiration.

The text's economic paradigm is institutional economics, also known as political economy, rather than neoclassical economics or Marxist economics. In terms of method, a theoempiric approach is taken, a cyclical movement between reflection and fieldwork, fieldwork and reflection. In addition, the philosophical distance from the natural sciences is stressed-in studying human society we explore the actions of purposive, reflexive agents and the families and institutions these individuals constitute. We may pursue useful theories but never universal laws.

The first proposal of this final chapter is that irrigation organisations as well as those that support them in various ways, such as MFIs and research and policy bodies, should appraise the potential of institutional economics in the policy area of water for agriculture. If their appraisal is positive, they should draw on this paradigm in their work.

The Demand for Irrigation Services

In understanding the institutional economics of irrigation we need to grasp the values, beliefs, motives and behavioural routines of individual farmers, as well as the irrigation sector's institutions. Farmers are enormously varied, of course, one from another. The simplest and most powerful distinction between them in any single cultivated region is likely to be between farming families and agribusinesses.

In output terms farmers who market their production seek crop tonnage, product quality and appropriate crop composition and timing of the harvest for the market. Where soil water availability sourced by precipitation to the rootzone of plants is insufficient in quantity or timing for crops' needs between planting and harvesting, then irrigation is required. The supply and use of irrigation water in agriculture demands that the economist consider the efficiency with which this natural resource is appropriated, delivered and applied. Irrigation water efficiencies can all be written as output-input ratios and the first to be explored in this book is *E*1, relative efficiency.

Here the input is measured as actual water supplied divided by the physical optimum for plant growth, and output is measured as actual crop tonnage divided by the physical maximum, with all other factors such as seed and fertilizer optimised. Relative efficiency is an agronomic concept and a tricky one at that because it is the ratio of two ratios.

It can be used to show, for a wide variety of crops, that a proportionate increase in water supply is matched by a similar proportionate increase in crop tonnage but that this response fades at higher levels of water supply. By implication there may be little agronomic or economic justification in seeking maximum plant yields. The second irrigation water efficiency is *E2*, requirements efficiency, the ratio of net irrigation requirements divided by the base supply.

Requirements efficiency is a distributional management concept. Assuming that a crop's net irrigation requirements are in fact met, it indicates the base supply actually used to resource those requirements. For example, at a value of 1.0 the whole of the base supply is taken up by crop evapotranspiration; at a value of 0.5 the base supply is twice the net irrigation requirements and one-half of the base supply flows off as drainage or is lost to non-beneficial evapotranspiration.

This measure can be of great use in 'crop per drop' discussions *provided* it is recognised that drainage water in one zone of an irrigation district is not necessarily lost to the district as a whole. The third irrigation water efficiency considered is *E3,* tonnage efficiency. This is a planning concept of limited interest, a ratio of crop output to the base supply of irrigation water. It is a close cousin of the farmer's concept of yield-output tonnage per hectare. Tonnage efficiency can be used in predicting the water resource requirements of the agricultural sector in a future year. In itself it is of little economic interest since it reflects neither output value nor input cost.

The irrigation authority levies no charge for the supply of water. Here the farmer regards irrigation water only as an agronomic input, not as an economic good. Farmers themselves are responsible for the irrigation water supply and bear the costs both of installing the infrastructure as well as operation, maintenance and management. The irrigation authority charges the farmer a volumetric price. The irrigation authority levies a charge for water such that the total charge payable varies *indirectly* with the volume used. The *warabundi* rotation of turns with water allocated in proportion to land is the best-known indirect charge. An irrigation service fee is levied that bears no relation to the volume of water used.

Only in cases 2, 3 and 4 above is the cost of water to the farmer of a type that resembles the pricing system for most market goods and services assumed by economists. Even the *warabundi* approach could not act as a price influencing the volume of water demand *per hectare* (although other indirect charging methods are able to do so). These reflections on charging systems and water prices lead on to consideration of the demand function for irrigation water.

This represents the quantity of water that would be purchased in a given market at each of a number of different prices. Economists measure the responsiveness of volume demanded to price change by the price elasticity of demand. This elasticity is defined, for any given price-quantity combination, as the proportionate change in quantity divided by the proportionate change in price.

The very important result obtained is that when price is low (and quantity purchased high) the value of the price elasticity of demand is low, of the order of "0.1. This implies that a 10 per cent increase in price would produce only a 1 per cent fall in quantity demanded. The farm budget is considered next; it is a tabulation of a farm's annual flow of expenditure and income. On the expenditure side, prime costs plus overhead costs are equal to total costs. Prime costs are defined as expenditure on those inputs used up in the daily round of farm work, including any spending on irrigation water. Overhead costs of production are non-prime costs and are often assumed to be invariant with the level of crop output.

The farm's gross margin per hectare is the excess of total income ('turnover') over total prime costs, divided by farm size, and is the source of income to cover the farm's overheads. Turnover less total costs equals the annual surplus of the farming family or agribusiness. The farmer's demand for water is itself derived from the market's demand for the farmer's outputs. In terms of economic theory the farmer may estimate for irrigated crop output a rather sophisticated water efficiency measure: the increase in farm tonnage following a modest addition to the base supply of irrigation water, i.e. *marginal tonnage efficiency* (*E*4). Fieldwork often suggests it has a zero value.

When the value is known either to farmers or to agricultural extension workers, it can be useful in moderating the demand for irrigation water.

Using their knowledge of how farm turnover would increase with the additional output and how the prime costs of water would rise, farmers would have a marginal efficiency measure *in economic terms (E5)*. *E*5 transposes *E*4 from physical to cash terms and can be called *marginal turnover efficiency*.

From this perspective, an increase in the price of water would probably be followed by a reduction in the demand for water. Similarly, other things being equal and for a known value of the net irrigation requirement, then the greater the gross margin of any single crop, the greater will be the quantity of irrigation water purchased for its cultivation at any given price of water. Conversely, when the farmer makes choices between the crops to be grown, the higher the price of water, the less likely it is that crops with low values of *E*3, *E*4 or *E*5 will be selected. Which one of these three efficiencies is used by farmers doubtless varies; in economic terms, *E*5 is the preferred criterion. Economic theory is by no means always close to farmer behaviour in this respect.

First, such rational action is most effective only in the cases of ownsupply cost and volumetric price.

Second, it is feasible only where the farmer has control over the volume of water he can apply to the land.

Third, the farmer may not know the response of crop output to marginal change in water use, adopting a satisficing rather than an optimising pattern of behaviour.

Fourth, if the price of water is low or if the proportion that water outlays make up in total prime costs is small, the marginal adjustment of this input may be seen as unimportant, so the elasticity of demand for irrigation water would be negligibly low.

Here, in respect of prime costs, irrigation water is not an economic commodity to the farmer but is simply an agronomic necessity. So in all catchments and for all families and agribusinesses, water is certainly seen as a vital natural resource in crop production, but is less often viewed as an economic input. In the latter case, the volume demanded will not be sensitive to modest price manipulations by a catchment agency. Reuse engineering is best interpreted as a supply-side innovation. At the same time, it brings about a lower aggregate demand for water by the community as a whole than would happen in the absence of reuse, and, in that sense, reuse is also treated as a form of demand management.

Second we have use technology. Water-use technologies that promise to achieve a higher level of requirements efficiency *(E2)* are sprinkler, drip and trickle irrigation. In practice, without adequate management, the new technologies can be less efficient than simpler surface methods.

The *third and fourth* components of demand management are land-use planning and environmental education.

The *fifth* component is water pricing, bringing us back to the demand for water in the strict economic sense of a price-quantity function. The metering of water use and setting a price per unit quantity help to underpin all the first four demand management forms.

Sixth and last, food import policies can play a very substantial part in management of the demand for water, particularly in semi-arid and arid countries. If a national government takes the decision to restrain the growth or cut-back the size of the agricultural sector and to place a growing reliance on imports to feed the domestic population, the demand for irrigation water will be less than in the no-policy-change situation.

Finally, the case studies introduced two more measures of efficiency. E6, gross margin efficiency, is the ratio for a farmer of his gross margin to the base supply of irrigation water. It failed to explain cropping patterns in the Curu Valley. *E*7, turnover efficiency, appeared in the Keveral Farm case study and is equal to the gross sales income of irrigated produce per cubic metre of irrigation water. It can be estimated at the farm, scheme and irrigation district levels. With respect to irrigation policy proposals, this is the place to address the issue of charges to farmers for the irrigation water with which they are supplied.

The second proposal of this final chapter is that where farmers already bear the full cash cost of supply-because they supply themselves-then government should level no additional charges. The exception is where an abstraction charge is required to finance the catchment agency or to moderate the volume of demand.

Own-supply irrigators excepted, *the third proposal* is that when all is for the best in the best of all possible worlds irrigation water should be volumetrically priced so that total charge income covers the total costs of government's or a user association's infrastructure, operation, maintenance and management. The users of this natural resource, those who benefit from its appropriation, should pay for the cost of supply, with the added advantage that the waste of the resource is reduced if not avoided completely.

Few citizens live in the best of all possible worlds, as Voltaire recognised (1759). Full-cost pricing may be technically costly in requiring accurate measurement of the flow of irrigation water to each farmer. Full-cost pricing may also be politically infeasible, particularly where farmers consider themselves disadvantaged by other national policies (such as subsidised grain imports) or where the urban élite corruptly escalates infrastructure costs to its own advantage. By way of illustration of this last point it is worthwhile pointing out that Clay Wescott, a senior Asian Development Bank (ADB) public administration specialist, said in 2001 that in the ADB region public expenditure was rent-skimmed on a large scale. In governance assessments already completed in Thailand and Vietnam the ADB had found that approximately 30 per cent of public expenditure in both countries 'vanished via fraud and corruption'.

In this case charges should be set at least to cover OM&M costs and paid directly to the irrigation supply authority, not to the regional or national Department of Finance.

If volumetric pricing is too ambitious for whatever reason, indirect fees should be explored based on irrigated area or crop type or time allocation. OM&M charges based on an indirect fee method is *the fourth proposal,* a fall-back solution where full-cost pricing is not feasible.

7

Watershed-based Integration of Hydrology

Binghamton University is located within the Fuller Hollow Creek watershed, an 8 km^2 basin draining to the Susquehanna River near the Pennsylvania border in southern New York. The headwaters are steeply sloped, primarily forested, and are sparsely developed. Downstream, the main branch of Fuller Hollow Creek is bordered by a suburban residential development to the east and by campus to the west. An 80 hectare Nature Preserve owned by the university borders the campus to its south, and includes forest, wetland, and a 2.4 hectare pond. In combination, the diverse characteristics of the watershed provide opportunities for interdisciplinary study of natural and anthropogenic processes within the confines of a university campus.

Interdisciplinary, watershed-based study was initiated at Binghamton University after creation of an undergraduate Environmental Geology Track within the Geological Sciences curriculum in 1998. This curriculum revision was based in part on pedagogical study that indicates that participating in a laboratory or field exercise enhances students' ability to relate to, analyse, and understand real world data and the process they are testing; and actively performing work rather than simply taking notes during lectures more than doubles students' comprehension and retention. We incorporated these pedagogical findings within a watershed-based focus during creation and revision of environmental geology courses to 1) promote active learning outside of the classroom, 2) allow students to work with state of the art field and laboratory equipment and analyse their own "real world" data, and 3) encourage interdisciplinary learning about the environment through use of the watershed on campus.

Our efforts to use the watershed as a teaching tool have been adapted and modified from experiences of others. The use of small well fields to provide a field laboratory for students of hydrogeology has been implemented with success in recent years. Others have coupled field investigations in their courses using various combinations of geochemical, geophysical, hydrological, and biological approaches [e.g. geophysics and geochemistry by Tibbs and Cwick (1994); hydrology, geochemistry, and biology by Panno et al. (1998) and Dunnivant et al. (1999); geochemistry and hydrology by Heins and Walker (1998), Carlson (1999), and Rimal and Stieglitz].

The target audience for these field experiences has varied from the introductory undergraduate level, the upper undergraduate level, and summer programs for K-12 students. Despite the varied approaches and audiences, these field-based educational opportunities share common benefits: that students relate to the subject matter better because it is local and "real world", have an enhanced sense of "ownership" since the data have been collected by them, (e.g. Dunnivant et al., 1999), and report more learning occurred in the field or lab than in the classroom.

Based on others' accomplishments, we determined that the short-term goal for this specific project was to enhance the undergraduate learning experience through the addition of significant field experiences on campus. Our long-term goal is not only to enhance students' classroom-based learning with field experiences but also to enable them to develop and conduct meaningful, field based, "capstone" projects at the undergraduate level. Our work has been supported in part by a grant from the National Science Foundation's Division of Undergraduate Education in 1999.

This project is different from many others in that we have installed a diverse variety of instrumentation for gathering environmental data at readily accessible field stations throughout the entire campus watershed. Included are wells, surface runoff weirs, pond stage gages, and a meteorological and atmospheric monitoring station.

In addition, the Geological Sciences Department has portable equipment for geophysical field studies (seismic refraction and reflection, resistivity, and gravity anomaly measurements) that can be used throughout the watershed. Using this instrumentation, the undergraduates are able to study the campus watershed from hydrogeological, geochemical, and geophysical perspectives and to monitor and explore differences in physical and chemical processes between and within the atmosphere, hydrosphere, and lithosphere on real-time scales. In this manuscript, we describe the campus watershed field stations, present examples of field exercises that others may implement at their institutions, and assess the impacts to date on our Environmental Geology curriculum.

Field Sites and Instrumentation

A well field was installed in the developed part of campus. It includes one well (approximately 36 m deep) terminated in Devonian shale bedrock and two wells screened in the overlying glacial till (approximately 6 m deep). Data on groundwater fluctuations with time is collected using pressure transducers and data loggers. A submersible pump powered by a portable generator is used to conduct pump and slug tests. Drilling spoils were collected during well installation and have been analysed by students to characterise variations in physical and chemical characteristics of the glacial till and underlying bedrock with depth.

This well field is adjacent to Fuller Hollow Creek, which students monitor for variations in flow and chemical composition. Data collection is ongoing at various points along the creek, including drainage from the headwaters and the Nature Preserve, and storm drainage inflows from the residential developments and from campus. Automated samplers are used for collecting water from wells, streams, and impervious surfaces for analysis. Students take samples during individual storm events and over longer time scales to assess causes for temporal variations in chemistry.

The second field site is the undeveloped campus Nature Preserve. Two monitoring wells were installed with piezometres screened at multiple depths: one at the till-bedrock interface at a depth of approximately 35 m, one screened to intersect the water table at approximately 6 m, and one intermediate between these depths at approximately 15 m.

These wells are instrumented with data loggers and pressure transducers to record groundwater hydrographs and are used to measure horizontal and vertical hydraulic gradients. Staff gauges are used to measure pond stage. Stream discharge from the ponds is measured manually. Portable monitoring equipment includes stream velocity metres, pH and conductivity gauges, ISCO automated samplers, a soil sampling kit, and infiltrometres. Geophysical equipment includes a Worden gravimetre for residual gravity surveys, a 24-channel Geometries Strataview seismograph for conducting shallow seismic refraction and reflection surveys, and equipment for collecting DC electrical resistivity profiles and soundings. The geophysical and portable monitoring equipment are used at Site 1 and 2 as well as other locations throughout the watershed.

Precipitation, temperature, relative humidity, barometric pressure, wind speed and direction are recorded at the meteorology station located at the Geology Department's building on campus. Evapouration is measured using a National Weather Service Class A evapouration pan located in the building's courtyard. An automated sampler collects rainfall and dry deposition between storm events for chemical analysis.

Research grade laboratory equipment used by the environmental geology undergraduates for analysing water and air samples includes a Perkin Elmer Elan 6000 ICP-Mass Spectrometre for trace metal analysis, an ARL DC Plasma Spectrometre for major element analysis, a Lachatt IC 5000 ion Chromatograph for ion analysis, a Tekran automated vapour phase mercury analyser, and an Electroscan 2020 Environmental Electron Microscope for aerosol particle imaging and analysis. Soil and rock samples are characterised using a Philips MPD X-ray diffractometre and a JEOL 8900 Electron Microprobe.

Use of the Watershed in Hydrology Courses

Environmental Hydrology is an inquiry-based undergraduate course at the junior (or 300) level, which alternates lectures with field-based ' laboratory" classes. At the beginning of the term, a question is posed to the students which relates to the hydrology of the campus watershed. For example, "How might construction of a new dormitory affect the hydrologie cycle in the Nature Preserve?" Individual "labs" address individual hydrologie processes or teach specific techniques, for example, determination of stream discharge or of evapouration. The students form small field teams to continue to collect data as (ungraded) "homework" assignments throughout the semester.

By semester's end, the data about the individual hydrologie processes from the labs and the field teams are integrated into a hydrologie report that addresses the question posed at the semester's beginning. While field work is done in teams, each student is responsible for preparing their own final report, as this is the main means of assessing their learning over the term. The hydraulic gradient indicates flow from Well 1 towards the ponds, but from the ponds towards Well 2. This type of data can lead to a discussion about groundwater recharge and discharge and surface water -groundwater interactions.

Use of the Watershed in Hydrogeology

Hydrogeology is a traditional, lecture-based class at the senior undergraduate (or 400) level that has been revised to incorporate field exercises and data analysis. Each class period is 1.5 hours long, limiting the time available for exercises (and making the accessibility of the field sites attractive). Hydrogeology students use data from the well field at Site 1 to examine temporal fluctuations in groundwater levels, and compare them with discharge in Fuller Hollow Creek and with precipitation records from the meteorology station at Site 3. While groundwater levels do fluctuate annually through wet and dry seasons, they do not reflect short-term fluctuations in response to storm events. The students conduct pump and slug tests at the well field to determine hydraulic properties of both the bedrock and the till. Students use handheld water level indicators to measure well drawdown and recovery with time during the aquifer tests. At Site 2 in the Nature Preserve, the students take manual measurements of depth to water in the multi-level piezometres, and then use their measurements to determine horizontal and vertical hydraulic gradients. Comparison with pond stage allows for inferences about groundwater interactions with surface water.

Use of the Watershed in Environmental Geophysics

Environmental Geophysics, a senior level undergraduate course, is an inquiry-based, field-oriented course in applied geophysics. Students use shallow geophysical surveying equipment to answer questions about the subsurface structure, which strongly influences groundwater flow and surface environmental conditions. For example, prior to well installation at Site 1, students used seismic refraction and electrical resistivity methods to predict depth to bedrock and depth to the water table.

In a subsequent year, the question was whether the knolls separating the Nature Preserve ponds from the lower portion of campus (near Site 2) were bedrock features or simply thick sections of glacial till. Another class concerned themselves with determining subsurface structure in an area of planned dormitory construction. In the most recent class, students determined what happens to bedrock between the shallow exposures on the hills forming the southern boundary of the Nature Preserve and the greater depth of bedrock (determined by a prior year's class) under the knolls forming the Nature Preserve's northern boundary.

Classes consist of a weekly 4-hour field experiment, supplemented by a weekly 1-hour lecture on theoretical background, methodology, and interpretation. Homework exercises allow students to practice data processing and interpretation using synthetic or known data sets similar to the ones they will collect in the field. After an initial reconnaissance of the field area and discussion of the issues of interest, students decide as a group where and how to carry out the various experiments. Seismic refraction is usually the first experiment since it gives a quick overview of depth to bedrock and the water table. This section shows distinct arrivals within the vadose zone, the saturated zone and in bedrock, with the latter suggesting a down-dropped section of bedrock. Gravity measurements are made throughout the field season by groups of 2-3 students while others carry out different experiments. Initial results of seismic refraction and gravity interpretations lead students to decide where to focus more detailed experiments with seismic reflection or DC electrical resistivity. After about two months of field work, the students have collected sufficient data so that remaining weeks can be spent processing and analysing data in the computer lab, using PCs and UNIX workstations. Continuing to work in small groups, they develop and orally summarise poster presentations of their preliminary results, testing, refining, and debating alternate hypotheses.

Although the data are collected and shared by the entire class, each group makes different decisions in data processing and analysis. Each student then writes an individual final report, patterned after professional reports by practicing Environmental Geophysicists.

Use of the Watershed in Environmental Measurements

Environmental Measurements is a senior level course offered in the spring term and is designed to be a "capstone" course for undergraduates concentrating in Environmental Geology. It is a field-based laboratory class, with weekly 3 hour "field labs". The students in this class conduct inquiry-based research on the geochemical processes related to soil, air, and water quality across campus. An example of a hypothesis driven question posed to the students at the beginning of the term is "How does the campus power plant impact the Fuller Hollow Creek watershed?' Early in the semester each week's lab activity introduces another measurement or analytical technique that could be used to address the question posed. Data collected by previous Environmental Measurements classes as well as by the fall semester hydrology classes is also analysed. Student teams use this information to design and implement sample and data collection that takes place during the semester. The synthesis of the many types of geochemical and hydrological data collected to provide different parts of the answer to the environmental question posed at the semester's beginning is a key aspect of this course.

The students discerned a relationship between meteorological parameters, like wind direction and wind speed, and atmospheric mercury concentrations. This analysis provided evidence of atmospheric transport and deposition of pollutants from the coal burned at the campus power plant, located 315° to the northwest of our meteorology station, to the campus watershed.

The students used automated flow-paced samplers to collect stream samples during a storm event; the timing of sample collection is indicated on the x-axis. The students examined concentrations of a variety of constituents in relation to stream discharge during baseflow and storm flow periods, and used these to identify constituents associated with groundwater baseflow and those associated with overland or storm flow. Concentrations of two elements that are predominantly in dissolved form, Sr and Ba, have an inverse relation to runoff volume. Concentrations of two elements that are predominantly in particulate form, Al and Mn, have an inverse relation to total dissolved solids (TDS) content of stream water. For Sr and Ba, and TDS, in-stream concentrations are diluted as the storm hydrograph peaks, and return to fairly steady, pre-storm levels after the hydrograph returns to baseflow. This behaviour indicates groundwater baseflow as the likely source of Sr, Ba, and TDS, and this data can be used to perform a geochemical separation of the hydrograph into groundwater and storm flow components. Al and Mn, by comparison, show a large fraction to be present in particulate form (the difference between total and dissolved fractions), with highest particulate loading during peak runoff. This type of data collection creates opportunities to discuss physical transport mechanisms like particle-bound transport as well as the impact of changing conditions on element-specific metal speciation behaviour.

By the end of the semester, the students integrate meteorological, hydrologie, and geochemical data on a team basis to develop mass balance based chemical loadings at the watershed outlet. They use this method to assess the impact of campus activities, including the campus power plant and parking lot and roadway runoff, on water quality. Interpretation and synthesis of the data is then done individually within final reports.

Assessment

The synthesis exercises at the end of the classes, in which students must assimilate the entire semester's worth of field data into a coherent interpretation of some aspect of the watershed, are a key means for assessing learning. Rather than focus on mastery of independent topics during the term or on memorization for a final examination, the final projects force the students to analyse and question the data they have collected, and to compare and contrast data from different methods or processes. Having a research question (e.g. "What impact does the campus power plant have on Fuller Hollow Creek?") helps focus their analyses. End-of-the-semester surveys have been used to assess the impact of the individual classes on student learning and on student opinions about science and research.

The questionnaires are anonymous and are used to supplement the standard, computerised surveys used by the university. Feedback about the watershed-focus and the inquiry-based approach has been positive; students report that they enjoy the practical experience in the field, and gathering and working with their own data. In each inquiry-based course, students wish they had more time at the end of the semester for data analysis and interpretation. Feedback in the Hydrogeology course, in which a full "inquiry-based" approach has not been incorporated, consistently includes "do more field work".

Experience by the instructors and constructive feedback from the students has helped improve individual field exercises from year to year. Originally unanticipated was the need to develop basic skills, like use of computer spreadsheets and simple statistics, to aid in data analysis. Most of the field exercises and data collection outside of class is done in teams; student feedback consistently reports student discomfort in working with teams due to issues with equity of workload. Concerns about receiving one "team" grade have been addressed somewhat by having teams responsible for data collection but individuals responsible for the final data analysis and reports.

Our self-evaluation as faculty indicates three areas in which we need to make programmatic improvements. First, we need to develop a means to better assess content learning in these courses and the longer-term impact of these field exercises and the "watershed focused" environmental geology curriculum on our students. Consistent with findings by other instructors at other institutions, our students report that they have learned more in these courses because of the field work than they would have in a lecture based class. To objectively assess this learning outcome, we have begun to implement pre-course questionnaires to assess the incoming students' knowledge and (perhaps) misconceptions. These will be compared with end-of-the semester assessments to better quantify student growth over the semester.

Second, we need to integrate the watershed-based focus earlier in our curriculum to foster student interest and provide research skills so that a greater number of undergraduates undertake independent research projects in their senior year. While we have had an average of one undergraduate student each year complete independent research projects using the watershed and monitoring instruments, we have not yet seen the growth in undergraduate independent research that we would like. Three of the four courses which have been most impacted by the Environmental Field Station are at the senior undergraduate (400) level. While these courses may generate student interest in environmental science and research, this is too late for an undergraduate to then begin a research project.

The two lead authors of this manuscript developed the current program while first-year faculty at Binghamton. As such, we developed a common vision for the watershed-based curriculum and a common plan for instrumentation of the watershed. However, we chose to first develop and implement field exercises into our existing, individual courses. This manuscript documents the results of that first phase of curriculum reform. We are now addressing the next phase in a planned curriculum revision to introduce the watershed focus and field exercises into the curriculum at the sophomore and junior level. The goal is to foster interest and provide research skills earlier in the undergraduate program to better enable more students to undertake independent research in their senior year.

Lastly, incorporation of an ecological component would broaden the present emphasis on physical and chemical processes. The current suite of classes serves the Environmental Geology undergraduates in learning about hydrology, geochemistry, and geophysics. Addition of ecology will require cooperation across departmental boundaries to include Biology faculty who participate in the interdisciplinary Environmental Studies Program. It will provide a more complete approach to the study of the watershed and will be addressed in future curriculum improvement proposals.

Transferability

We recognise that not every university environment has opportunities for hands-on watershed-based activities within walking distance from the classroom. However, watershed-based study of environmental geology can be adapted to many settings, even the most urbanised. We believe some or the concepts or exercises described in this manuscript could be used as a template on an individual course or integrated curriculum basis elsewhere. Funding for monitoring equipment is obviously helpful, but lower cost alternatives could be utilised.

For example, monitoring wells at a local landfill could provide access for mapping the water table, or a partnership with a local well driller could provide access for well testing. USGS river gauges could be used to examine temporal fluctuations in surface water, and to provide hydrographe for analysis of froundwater discharge and recharge in larger river asins. Hand held pH and conductivity probes, together with a stream velocity metre and staff gauge can be used for basic stream analyses. Low cost, portable geochemical assessment kits, while not suitable for research grade geochemical analyses, can serve for field-based stream assessments. Incremental implementation, starting with pedagogical change in individual courses, is advisable. As the individual courses develop, and along with them a common faculty value of the importance of "student centreed" learning, implementing such change across a whole curriculum can be approached. A major factor to be considered before implementation of watershed-based study across multiple courses (for example, in both spring and fall course offerings to capture most of the annual hydrologie record) is the need for ongoing cooperation and coordination between faculty members.

Readily accessible field sites, instrumented with state of the art equipment, have been established on Binghamton University's campus. This has allowed us to turn our campus watershed into a "field laboratory" for use by the undergraduate Environmental Geology students. This watershed focus and field instrumentation has had a major impact on four undergraduate courses: Hydrology, Hydrogeology, Environmental Geophysics, and Environmental Measurements. Through this project we have met our goals for promoting active learning, giving students experience in collecting and analysing their own "real world" data, and promoting interdisciplinary thinking. The students have gained an appreciation for the complexity and inter-relationships of watershed-based processes.

Watershed-based Summary Field Exercise

During the past decade, field coursework in hydrogeology has proliferated within undergraduate and graduate Earth science curricula. An informal Web search found at least 10 institutions offering summer field courses in hydrogeology or hydrology as of 2005, and other institutions have integrated hydrogeology modules into conventional geology field camps. Some Earth science programs have instrumented watersheds or well fields on or adjoining their campuses, and have integrated data collection and interpretation from these sites into regular coursework. As noted by those authors, field exercises promote comprehension and retention of fundamental concepts, and they introduce students to techniques used by professional hydrogeologists. Moreover, placing such exercises within a watershed framework facilitates understanding of the linkages between surface and subsurface hydrologie processes that are commonly decoupled in discussions of water resources. However, some students are not able to take summer field courses, and "not every university environment has opportunities for hands-on watershedbased activities within walking distance from the classroom".

In this chapter, we review a watershed-based, summary field exercise used in an introductory hydrogeology course (GLY 585) at the University of Kentucky (UK). GLY 585 is a 3-credit hour lecture course that is typically offered during the spring semester and is taken by both undergraduate and graduate students. Total enrollment for each class ranged from 5 to 20 (average 13, median 12) between 1996 and 2009. Geology students have been a majority in most classes. However, because the only prerequisites are physical geology and first-semester calculus, students from other sciences, engineering, agriculture, and science education also have enrolled. Student performance is evaluated by midterm and final exams, several problem sets, a term paper, and a presentation, which is optional for undergraduates. During the first four times the senior author taught the course, field trips were 1-day tours of local karst features and clastic (alluvial) aquifers farther away. In 2000, we began a weekend trip as described below, which is analogous to trips for other upper-level geology courses at UK but with inclusion of numerical data collection and analysis.

Study Area Setting

The Ledbetter Creek watershed covers 24 km2 in Galloway and Marshall counties in the Gulf Coastal Plain of western Kentucky. The area features rolling land between sharply incised valleys, with land-surface elevations ranging from ~ 158 m above mean sea level (amsl) along the divide to ~ 109 m amsl at the mouth of the watershed. Cretaceous to Holocene sediments unconformably overlie Mississippian bedrock and dip gently westward. Land use in the watershed includes farms, rural residences, and parkland. Land cover is ~ 62% forest, 17% grassland, 15% cropland, 5% brush, and 1% water.

The region has a humid temperate climate, with moderately cold winters, warm summers, and no distinct wet or dry season. For the period 1971-2000 at Murray, Kentucky (~ 20 km southwest of the Ledbetter Creek watershed), air temperature ranged from an average minimum of -2.8 0C in January to an average maximum of 32.2 0C in July, and annual precipitation averaged 140.4 cm.

Ledbetter Creek is a third-order perennial stream that drains to Kentucky Lake, the terminal reservoir on the Tennessee River that was formed when the Tennessee Valley Authority (TVA) constructed Kentucky Dam in 1945. Current TVA management practice is to raise the reservoir level ~ 2 m in March to summer pool, then lower it back to winter pool over a 3-month period beginning in August. Groundwater flows regionally toward Kentucky Lake and locally toward Ledbetter Creek and its embayment (the perennially flooded lower reach of the creek, which is an arm of the reservoir). Fluctuations in reservoir stage can cause temporary reversals in hydraulic grathent adjacent to Ledbetter embayment.

Activities

Preparation for the summary exercise begins in the first half of the course, with discussions of groundwater/ surface-water interactions more detailed than those in the textbook. Since 2003, these discussions have been tied to two problem sets involving data compiled by the senior author from elsewhere in the region. The first assignment entails hydrologie balance calculations for a lake in the Ohio River floodplain during 1998. These include using: (1) climatic data to calculate the volume of precipitation falling on the lake and estimate evapouration from the lake during the monitoring period; (2) gauging data with an Excel spreadsheet to calculate discharge from a channel draining the lake; (3) Darcy's law to estimate seepage out of the lake; and (4) results of the preceding calculations to estimate seepage into the lake.

The second assignment includes hydrograph separation using daily streamflow data from a U.S. Geological Survey (USGS) gauging station in the region; calculating vertical hydraulic grathents for a nest of monitoring wells in an alluvial aquifer along the Ohio River; and mapping hydraulic heads in that aquifer. The data in this last problem were collected in September 1997, when the river's stage was relatively low. For comparison, to reinforce the concept of bank storage, students are subsequently shown the Ohio River hydrograph and hydraulic heads in monitoring wells from March 1997, when the largest flood in 47 years occurred. As a result, hydraulic heads were elevated and lateral hydraulic grathents were virtually stagnant as far as 5 km from the river.

The summary trip is usually scheduled during a weekend in the latter half of April (i.e., during the last two weeks of the semester). Prior to departure, students are given an overview lecture on the study area and provided with maps and elevation data. The class is based during the exercise at Murray

State University's Hancock Biological Station (HBS), which is located in an adjoining watershed that drains to Kentucky Lake, approximately a 5-hour drive from UK. Field activities typically take one day and focus on flow and water-level measurements.

At two locations on Ledbetter Creek, we gauge stream flow by wading using the midsection method with an open-reel fiberglass measuring tape, top-setting rods, and a digital flow metre. We measure seepage from the stream bed at gauging sites using a homemade seepage metre. A flattened condom is pulled over the larger rubber stopper and allowed to fill during a timed interval, after which the condom is pinched and removed, and the volume of water collected is measured in a 100-mL graduated cylinder. We also measure discharge of a small perennial spring flowing into the embayment using a bucket, stopwatch, and 2-L graduated cylinder.

Water-level measurements include observing an artesian monitoring well on the opposite side of the embayment from the spring. This well consists of 2-inch nominal (5.3-cm actual ID) PVC pipe, which was installed by a licensed driller with a truckmounted hollow-stem auger rig through alluvium and into bedrock, to a depth of 12.2 m below ground level (bgl). The well is screened from 10.7 to 12.2 m bgl. Using an electric water-level indicator, we measure depths to water in several sets of piezometres in the alluvium. These consist of sections of 4-inch nominal PVC pipe, which were slotted or perforated along an interval of ~ 0.3 m, capped at the bottom, and emplaced in holes drilled with a gasoline-powered auger rig operated by two persons. The piezometres include two in a transect across the Ledbetter Creek floodplain (3 m bgl), two along the axis of the embayment (3 m bgl), and five nests (each containing two to three piezometres completed to depths of 0.6 to 3 m bgl) across the embayment. With a folding rule, we measure surface-water levels relative to the top of casing for the embayment piezometres, which extend ~ 2 m above ground surface. Because the depth of water in the embayment can be > 1 m in April, we access offshore piezometres by wading or via pontoon boats.

Lastly, we measure casing heights and depths to water for seven domestic wells around the Ledbetter Creek watershed. Six of these wells are amenable to water-level measurements because they were constructed of 2-ft (0.6-m) diametre concrete culvert pipe (a local driller used a bucket auger to excavate gravels and cobbles). Because of the relatively large casing diametre, the electric tape does not become entangled around the smaller diametre PVC water-supply pipe or submersiblepump wiring within the well. The remaining well (#3), which is not in use, consists of PVC pipe without a pump installed. Depths to water vary from ~ 10 to 35 m bgl. One of the larger-diametre wells (#6) is perched, as indicated by water-level measurements ~ 9 to 12 m shallower than those in an adjoining deeper well (#7). To prevent crosscontamination between domestic wells, the bottom ~ 0.5 m of the electric tape is rinsed with bleach and distilled water after each measurement. Equipment and supplies used in the exercise are listed in an online supplement.

Data Analysis and Assessment

Following field work, the class meets to ensure that everyone has recorded the same data. Within 2 weeks after returning to UK, the class is given a set of multi-part questions that constitute half of the final exam (the takehome component). The format of the take-home has remained the same since 2003. In the first question, students are asked to (a) modify the Excel spreadsheet from the first problem set to calculate gauged discharge (in m^3/s) at the upstream and downstream sites along Ledbetter Creek, (b) determine the net discharge AQ between the sites, (c) estimate the distance t between the sites from the topographic map (~ 686 m [2250 ft]), (d) calculate an average width w for the gauged transects, and thus (e) estimate specific discharge q for the stream reach between the sites (q = 94Q/ A, where the crosssectional area A = £w). Students are then given the crosssectional area for the seepage metre (0.0079 m^2) in order to calculate specific discharge for the seepage metre at each site, and asked to compare the gauged and metred values. In the second question, students are given approximate land surface elevations for domestic wells and, using their field data and the topographic map, asked to draw equipotentials and flow lines for the watershed. (The question does not refer to either a potentiometricsurface map or a water-table map because the main aquifer occurs within different strata and probably is semi -confined.) In the third (last) question, students are asked to calculate the vertical hydraulic grathent between various embayment piezometres and the lake, then asked whether the observed directions of flow are qualitatively consistent with the presence of the spring along the embayment and with their answers to questions 1 and 2.

Apart from minor errors in arithmetic, unit conversions, and data entry into the spreadsheet, students typically obtain reasonable values of specific discharge. Net discharge is positive between the gauging locations, which is assumed to reflect baseflow. As students are told in the field, inflow into the seepage metre may be affected by stream-water advection through the bed or elastic response of the condom. Although the values of specific discharge measured using different techniques are commonly within an order of magnitude of each other, students tend to recognise that the values obtained by gauging and by seepage metres represent different scales of measurement.

Mapping groundwater flow in the watershed is constrained by the paucity of data points and the fact that well #1 is actually located ~ 500 m beyond the divide. The perched well is excluded from contouring. Students are instructed to use the topographic map to infer hydraulic heads where wells are absent and to assume isotropie conditions, so that flow lines are perpendicular to equipotentials. For consistency with the topographic map, equipotentials are contoured in feet rather than in metres. Most students calculate hydraulic heads properly, correcting for casing stick-up (which is actually within the precision of the land-surface elevations estimated from the topographic map), and they draw flowlines away from the watershed divide and toward the embayment (where hydraulic head for the artesian well is assumed to be at land surface). Errors result where flow lines are not drawn converging toward the creek as well as the embayment, and where equipotentials do not approximate the topography of the watershed (e.g., because Ledbetter Creek is a gaining stream, hydraulic heads should be approximately equal to land-surface elevations along the creek). Calculating the vertical hydraulic grathent (iz) between groundwater and surface water is mathematically simple but conceptually subtle. Students are given a diagram in the lecture notes during the first half of the course illustrating that the hydraulic head of a stream or lake, similar to that of groundwater, is merely the water-level elevation above some datum, and the distance L over which the head drop occurs is the depth of the well or piezometre intake below the stream or lake bed. For simplicity in the final problem, students are told to assume that the intake is at the bottom of the piezometre. Students typically calculate the head difference correctly, but some forget to divide by L. In late April, vertical hydraulic grathents are often near zero or downward in the shallow piezometres and upward in the deeper piezometres near the axis of the embayment. As some students note, upward hydraulic grathents are consistent with other evidence of groundwater flow toward the embayment. However, only a few have recognised that near-zero or downward hydraulic grathents are probably a temporary artifact of recent reservoir-level rise, analogous to bank storage.

Each take-home exam is scored out of 50 points and returned to the student after the semester. For 2003-09 classes, minimum values ranged from 18 to 30, maximum values from 43 to 49, and medians from 36 to 43. We examined tendencies in scores on 2003-09 take-home exams by level of education (undergraduate [UG] versus graduate/ post-baccalaureate [G/ PB]) and by discipline (geology majors [undergraduate and graduate] versus others). Minimum, maximum, and median scores were greater for G/ PB students than for UG students in most classes. Maximum and median scores for geology majors were greater than values for other students in all but two instances. However, in four out of six classes, minimum scores for geology majors were less than for other students. Maximum and median scores for the four populations in the 2003-09 classes fluctuated within 12-point ranges, thus suggesting that student performance and grading have been reasonably consistent over time. Introduction of the summary exercise coincided with a general improvement in student ratings on standardised, end-of-semester teacher-course evaluations (TCEs; UK Office of Institutional Research, Planning, and Effectiveness [OIRPE], 2010). We compared ratings for classes prior to the exercise (in 1996, 1997 [spring and summer], and 1998) with those from 2000 through 2009 (note GLY 585 was not offered in 1999 and 2007). Averaged minimum, maximum, and median scores increased for five key items:

- Increased my ability to analyse and evaluate;
- Course helped ability to solve problems;
- Gained understanding of concepts and principles;
- Overall value of the course;
- Overall quality of teaching.

However, the increased scores cannot be attributed solely to the summary exercise. Other changes included the textbook (Domenico and Schwartz [1990] during 199698, Domenico and Schwartz [1998] during 2000-02, and Schwartz and Zhang [2003] from 2003 onward); lecture delivery (use of a blackboard in 1996 and 1998, a document camera in spring 1997, video broadcast in summer 1997, overheads from 2000 through 2002, and PowerPoint with handouts plus a whiteboard from 2003 onward); and introduction of term-paper presentations in 2002 and the western Kentucky problem sets in 2003. Separating the TCE responses further into 2000-02 and 2003-09 cohorts, we saw an increase in median scores coinciding with post-2002 instructional changes, although there were fluctuations in maximum and minimum scores with time.

To obtain feedback specifically on the summary exercise, we sent a questionnaire to 71 former students (out of 96 who took GLY 585 for credit between 2000 and 2006) for whom we had valid e-mail addresses in summer 2007. Forty of those students replied. Anonymity was maintained to the extent possible by not asking when the respondent took the class and by requesting that questionnaires be returned to the department's administrative assistant, who removed identifying information before printing the responses. However, several former students responded directly to the senior author. Slightly more former graduate students than undergraduates replied, and most replies came from geology majors. Of the 15 respondents who had subsequently been employed in hydrogeology, seven had not pursued further studies in hydrogeology beyond GLY 585. All of those who had attended the field trip (39 of 40 respondents) said it was worthwhile, and 36 of 40 said the field exercise and previous class activities gave adequate preparation for the take-home exam. When asked the most and least useful parts of the exercise, 12 respondents said all were most useful and 19 said none were least useful; the most common recommended change was "none". Among respondents who specified one or more techniques as being most useful, the most common answer was hydraulic-head measurements in piezometres and wells. Responses about useful parts of the exercise did not vary systematically by level of education (UG versus G/ PB), discipline (geology majors versus others), or having experience in hydrogeology after GLY 585 (coursework, research, or employment). Most additional comments were positive. Criticisms included the need for more time in the field, visits to other field sites, and more measurements by students.

Instructional Effectiveness and Limitations

Assessing the effectiveness of the summary exercise is complicated by the nature of the available data. Exam scores do not provide information about performance on individual questions and TCE scores do not specifically address the summary exercise. However, since 2003, ranges of take-home and TCE scores have been relatively consistent and TCE scores have usually exceeded college means. We therefore infer that the instructional approach underpinning the exercise is reliable. According to Wiggins and McTighe (2005), the validity of an exercise hinges on student understanding of concepts and not simply recall of formulas. In this case, the tendencies of students to recognise differences in scales of groundwater discharge (question 1), to infer directions of groundwater flow from contouring hydraulic heads (question 2), and to articulate plausible explanations for vertical hydraulic-head drops (question 3) represent a fundamental level of conceptual understanding. Moreover, survey comments indicate that former students found the exercise worthwhile not merely because "the performance is complex and the task interesting", but also because it was useful, practical, and like a professional project.

Notwithstanding the apparent reliability and validity of the exercise, there are several limitations and opportunities for enhancement. In part, these stem from the limited time available for the exercise, which is a consequence of the site's distance from campus. Additional time would facilitate activities such as installing new piezometres, surveying their elevations, conducting pumping tests, and sampling water for chemical analyses. Because of limited time and equipment, the instructor typically performs stream gauging while students record data. In addition, estimates of groundwater inflow have assumed that runoff is negligible between gauging sites. We think this assumption is tenable because rainfall at the nearest National Weather Service station (Paducah, Kentucky, ~ 66 km west-northwest) was < 1.0 cm for at least 24 hours prior to each time Ledbetter Creek was gauged. However, interpretation of gauging results following heavier rainfall should account for the possibility that the stream was not at baseflow.

The summary exercise described in this chapter addresses concepts and skills, including measurement of hydraulic heads and mapping of groundwater flow, that are fundamental to hydrogeologic research and professional practice. In the process, the exercise focuses on groundwater/ surf ace-water interactions, a commonly underemphasized topic in hydrogeology classes. Students engage in hands-on collaborative learning through field data collection and interact with local stakeholders (residents whose wells are monitored). Important quantitative skills utilised in data interpretation include unit conversion, dimensional analysis, and calculation of grathent.

Moreover, students can observe effects of reservoir management on groundwater flow. Since the 1940s, reservoirs have proliferated across much of the USA, yet standard hydrogeology texts do not usually consider them (one exception is freeze and Cherry [1979]). Although students have not been asked to compare their results with those obtained by previous classes, the opportunity exists to assess year-to-year variability in parameters such as stream flow, spring discharge, and hydraulic heads.

The techniques and materials used in this exercise are broadly transferable to other times of year and other locations, including natural lakes. Less expensive alternatives exist for some equipment (e.g., using a hand auger instead of a gasoline-powered rig to drill holes for piezometres). Having both a spring and existing wells available for monitoring within the Ledbetter Creek watershed, as well as a nearby fixed-base camp, is serendipitous but not unique. Use of a chalked steel tape, rather than an electric tape, should be considered to avoid entanglement during hydraulic head measurements in standard diametre (e.g., 4-inch nominal) domestic wells.

8

Water Quality Testing of High-Use

Recent media reports have focused on potential adverse health effects of local health clubs and spas. The Cuyahoga County Board of Health (CCBH) responded to these reports with a comprehensive review of sanitary and environmental conditions of these facilities including locker rooms, pools, and equipment. Spas were placed under the greatest scrutiny because of their potential adverse effects on public health and safety. Although public bathing places are currently inspected and are not considered to be a major public health problem, the study focused on the water quality of spas because of: 1) the potential for transmission of intestinal diseases and infections of the skin and mucous membranes, 2) large bather loads in heavy-use periods, and 3) the existence of measurable health-code standards.

Studies conducted by the German Federal Health Office and the Oregon State Health Department suggest that the parts per million (ppm) of disinfectant alone is not an accurate measure of bacteriological water quality unless other factors such as pH, total dissolved solids (TDS), and cyanuric acid are all maintained at acceptable levels.

As a more reliable alternative, a direct measurement of the oxidation-reduction potential (ORP) has been proposed by the Ohio Department of Health to monitor the water quality of spas and will be required statewide by January 1, 1996. In light of these developments, CCBH initiated a study to combine the standard water chemistry testing with bacteriological sampling to determine the efficacy of disinfection treatment, the confidence in standard chemical testing, and ultimately, a more accurate profile of the quality of spa water.

Materials and Methods

The preliminary study was designed within the framework of health code regulations and based on the protocol from Standard Methods for the Examination of Water & Wastewater. The bacteriological standard from Public Swimming Pool, Spa, and Special Use Pool Rules is, "Not more than two hundred bacteria per millilitre as determined by the standard (thirty-five degrees Celsius) agar plate count. A pool or spa shall be in violation of the standards if more than fifteen percent of at least four samples taken within thirty days exceed one of the standards". Because coliform bacteria are insufficient to judge the microbiological quality of whirlpool water, the heterotrophic plate count, formerly known as the standard plate count, is the primary indicator of disinfection efficiency for indoor pools. The membrane filter method, in particular, permits the testing of low turbidity water and is the method of choice for low-count waters.

While this study was designed to meet the above guidelines, it was limited in scope by: 1) laboratory capacity, 2) incubation time, and 3) collection and processing time. The available space allowed 48 plates to be incubated for a period of 48 hours. Since each sample yields three plates, only sixteen spas could be sampled in any 48-hour interval. Furthermore, the time required for collection and processing necessitated the use of two sanitarians (each collecting a maximum of eight samples per 48 hours) so that all samples were delivered to the lab and processed within eight hours. One final restriction, the 48-hour incubation period, precluded sampling on Thursdays and Fridays. The finalised schedule allowed for the collection within a thirty-day period of five samples each from a total of seventeen spas at eight east side and nine west side locations. Spas in the health/racquet/exercise club category were selected due to their heavy use and potential for producing a worst-case scenario relative to other spas. The sampling at these establishments was unannounced and occurred at different days and times throughout the study period.

The sampling protocol was adhered to closely in order to minimise any variation in technique or procedure between the two project sanitarians. During the initial visit to each establishment, the program manager of the pool program accompanied the project sanitarians to survey the pump room and spa facilities, explain the purpose of the study to the operators, and evaluate the technique of the sampling procedure. The established procedure required the sampling sanitarian to:

1. Note the number of bathers and ask them to exit the spa.
2. Turn off the jets and allow the water to remain calm for five minutes prior to obtaining a sample for bacteriological analysis.
3. Conduct water chemistry tests, (temperature, free and total chlorine, pH, TDS, and cyanuric acid), with a Taylor K-2005 test kit.

4. Apply an alcohol preparation to both hands before putting on one disposable latex glove.
5. Collect 100 mL of spa water into a sterile Whirl-Pak Thio-Bag. [The bag, which contains sodium thiosulfate to neutralize the disinfectant, is held inverted by the gloved hand into the water one foot deep and one foot out away from skimmers or return lines.]
6. Store the sample on ice and transport it to the lab within six hours.
7. Process samples within two hours using the HPC Membrane Filter Method at dilutions of 10, 20 and 40 mL with m-HPC media at 48-hour incubations.

Plated samples were placed in a Thermolyne culture incubator with a dish of distilled water to maintain the humidity within the unit. Upon completion of the incubation, the colony-forming units (CFU) were counted with the aid of a stereoscopic microscope at a 10x to 15x magnification and a light source which was placed at a 45-degree angle to the plates. The procedure for calculating CFU is based on the number of colony-forming units per square as follows:

1 to 2 or fewer CFU per square.

Count all colonies on the plate & multiply by the reciprocal of the dilution.

3 to 10 CFU per square.

Count ten squares, obtain an average count per square, and multiply by 185 times the reciprocal of the dilution.

10 to 20 CFU per square.

Count five squares and multiply the average count by 185 times the reciprocal of the dilution.

The seventeen spas involved with this study may be further divided into three classes: (A) franchised health club, (B) racquet club, and (C) fitness centre. Class A consists of eight establishments representing two different franchises. Class B is composed of three establishments with clientele whose primary activity is tennis, racquetball, or squash. Class C is represented by five establishments (one with separate men's & women's spas) with general fitness equipment and activities.

A total of 81 samples were collected from 17 spas at 16 establishments (four scheduled samples could not be collected because the spas in question were closed for repair). Four of the spas were equipped with automatic controllers which read disinfectant levels and adjust the amount of disinfectant fed to the spa water accordingly. Disinfectants used at these spas include sodium hypochlorite (11 spas), bromine (5 spas), and stabilised chlorine (1 spa).

The Ohio state health regulations require spa water to have a free disinfectant residual of at least two parts per million (ppm), a pH between

7.2 and 7.8, a cyanuric acid concentration less than 100 ppm, a total dissolved solids concentration less than 3000 ppm, and a heterotrophic plate count less than 200 CFU/ mL.

The corresponding water chemistry tests identified: free chlorine/bromine violations in 17.3% of the samples (14/81), pH violations in 14.8% of the samples (12/81), cyanuric acid violations in 0% of samples (0/4) from spas using stabilised chlorine, TDS violations in 6.2% of the samples (5/81), and 4.9% (4/81) of the samples exceeding the bacteriological standards. The letters A, B, and C refer to the three classes: franchised health club, racquet club, and fitness centre, respectively. The fifth column shows the percentage of samples from each spa which exceeds the bacteriological standard while columns six through nine do the same for the chemical parameters (disinfectant, pH, TDS, and cyanuric acid).

The minimum allowable concentration of chlorine or bromine is 2.0 ppm. The acceptable range for pH is 7.2 to 7.8. The TDS units depicted on the bar graph must be multiplied by 100 to calculate the actual values for total dissolved solids. The HPC is an extremely sensitive test which is much better suited than the coliform test for judging the efficiency of water disinfection for several reasons. Because coliform bacteria are naturally abundant in the intestines of warm-blooded animals, this group of bacteria is an excellent indicator of sewage pollution. In the spa environment, however, feces is only one type of human waste product which may be present in the water. The heterotrophic bacteria are ubiquitous and, because they are found on the skin of humans, may easily be introduced into spa water. This fact highlights the need for extra precautions to prevent contamination during the sampling and processing procedures.

Under ideal conditions, the bacteriological samples should be taken during peak use periods. Based on the observations of local spa operators, the busiest season is during the colder months when people tend to exercise indoors, with the peak time being late afternoon or evening. Most of these samples were taken before 2:00 p.m. during the months of May and early June. Even though the sampling events were unannounced, it is important to consider to what degree the spa operators' anticipation of additional sampling may have biased this study.

This study was designed to test the premise that, given an adequate concentration of disinfectant in chemically balanced water (proper pH, TDS, and cyanuric acid), analysis for heterotrophic bacteria should yield very low counts - well below the health code standard. This in fact is known to be true if the majority of free chlorine is in the form of hypochlorous acid, the component with a faster bacterial kill time. Since chemical test kits cannot isolate and measure the concentration of this one component of free chlorine, it is necessary to monitor other factors such as pH, TDS, and cyanuric acid which when not within ideal parameters can decrease the percentage of hypochlorous acid. In order to maximise the percentage of hypochlorous acid, the pH must be 7.4 to 7.6, the TDS below 2000 ppm, and cyanuric acid concentration below 100 ppm. The pH level has added significance because free chlorine is most effective as a disinfectant at levels lower than the limit established in the health code.

This lower limit protects bathers from eye irritation associated with pH levels below 7.2. The results of this study indicate that 92.5% of the samples with water chemistry parameters within the above range were in compliance with the bacteriological standard. It is interesting to note that all four samples with the number of CFU exceeding the bacteriological standard come from spas which use bromine as a disinfectant. Unlike chlorine, bromine is more effective as a disinfectant when the pH is greater than 7.8, the upper limit set forth in the Ohio health code. Furthermore, levels of bromine twice the amount of chlorine have been recommended to ensure adequate disinfection. Based on this information, those samples (having at least 200 CFU/mL) with a pH higher than 7.8 and bromine concentrations greater than 4.0 ppm may be false high counts resulting from either sampling or analytical error. The inadvertent capture of floating debris, for example, is an example of an error which might occur while collecting a water sample.

Recommendations

Operators of spas are responsible for maintaining equipment and facilities, monitoring bathers, and supervising employees so that they provide a safe and healthy environment at all times. Recommendations which will facilitate the maintenance of good water quality require the operator to:

1. operate only approved filtration equipment of adequate design and capacity.
2. maintain proper levels of disinfectant, pH, TDS, and cyanuric acid through frequent testing and required documentation of the water chemistry testing.
3. properly calibrate automatic controllers at the start of each day of spa operation.
4. monitor the entry of bathers to ensure adequate hygiene.
5. maintain a smooth and easily cleanable interior spa surface to eliminate areas which may harbour, and promote the growth of, bacteria.
6. frequently clean and disinfect floor surfaces leading to the spa area to reduce tracking of foreign objects into the water.

The results of this preliminary study support the use of chemical testing as an effective means for routine monitoring of spa water quality. This method provides reliable and immediate water quality data which facilitates the identification and correction of conditions which may pose a threat to the safety and health of bathers. Additional studies should be conducted during peak season and time of day as well as at spas other than those located at health, racquet, and exercise clubs.

Organisation Theory: Analysis of Water Management

The narrative of progress has been central to modern organisation theory, but as other narratives have come to the fore in contemporary organising scholarship, it is now being challenged. Using narrative analysis, this chapter investigate's contemporary water management in England and Wales. Four key narratives are identified, and these are shown to present a number of challenges to modern organisation theory. Contemporary organisational theorists from a variety of perspectives have sought to overcome these challenges, but we argue that they have not succeeded. Instead, we propose an alternative way forward-one that takes us beyond the current deadlock. This chapter argues that the narrative of progress is central to modern organisation theory, but that as other narratives have become dominant in contemporary organising, so modern organisation theory is being challenged. We present the results of a narrative analysis of contemporary water management in England and Wales. In doing so, we investigate the status of the narrative of progress and its relationship to the new narratives highlighted by the study, as well as the resulting implications for organisation theory. The chapter commences with a discussion of the narrative of progress and its importance for modern organisation theory. The next section explains the concept of narrative analysis and examines the relationship between the text of narratives and the content in which they are constructed and used. This is followed by a short overview of the water management industry in England and Wales.

The chapter then moves on to analyse four key narratives in water management that are identified by the study. The penultimate section comments on the challenges to modern organisation theory set by these narratives. The chapter concludes with a discussion of how contemporary organisation theorists from a variety of perspectives have responded to these challenges. It argues that, in the main, these responses have failed to advance organisation theory and that in order to make headway an alternative, narrative-based, approach is needed.

Modern Organisation Theory

In order to understand how new narratives might challenge modern organisation theory, we have to identify its defining features. Although not the first to be put forward, the theories of Max Weber (1997 [1948]) and Frederick Taylor (1947 [1911]), are generally considered to inform much of modern organisation theory and public administration in Western industrialised economies. There are profound differences between Weber and Taylor, but in the context of this chapter it is important to note that both view the emergence of the modem organisation as part of the rationalisation process in general. Rationalisation with regard to the organisation means in the first place that organisations are oriented toward efficiency, that is, the reduction of waste. It is accompanied by demystification of the world through the use of objective, scientific methods so that the organisation is freed from subjective and personal influences. This in turn means that the principles are universally applicable and are true for all times and all places. Consequently, an organisation can be designed in detail, in advance.

Although Weber intended his theory of bureaucracy to apply to every organisation, it was exemplified for him, above all, by the state bureaucracy. For a modem state, a bureaucracy is indispensable. He regarded the state as the central controlling and organising authority for society. Unlike Weber, Taylor did not focus on the organisation of the state, but his concept of scientific management as fully incorporated in modern theories of public administration. In particular, since Woodrow Wilson's plea for the separation of politics and state bureaucracy, similar organisation principles have been applied by organisational theorists to state and nonstate organisations. To take Weber's theory of bureaucracy and Taylor's scientific management as benchmarks for modem organisation theory is not to deny alternatives. In the first place, there were alternative theories available other than these.

Second, there have been major amendments since that time to the ideas embedded in both the American and the European origins of modem organisation theory. Nevertheless, many contemporary theories are still under the spell of these "old modern" perspectives, namely, those which are still engaged with rationality or the lack of rationality. Further, there has been a resurgence of modem rationality-based organisation theory. This, together with the complex interdependencies existing between theories, the deep assumptions they include, and the way they guide description of the world and recognition of key entities in the world requires both theoretical and empirical critique.

We cannot understand old modern theories of organisation without addressing the grand narrative at that time. Grand narratives concern the universality of historical accounts. and, as such, are efforts at universal understanding. At the time the theories of Weber and Taylor first took hold (late nineteenth century and early twentieth century, respectively), the grand narrative of progress shaped all spheres of Western society. This narrative is characterised by the line of reasoning that industrial society is more advanced than all previous stages in human history—and that the future is even brighter. Industrial society is urbanised, centralised, democratic, secular, but above all, rational and bureaucratic. The causes of success in industrial society are science and technological invention, in turn based on rationality and reason. There is a clear interplay between old modern organisation theory and this grand narrative. Scientific progress, so important in the narrative of progress, can be seen as the foundation of old modern organisation theory: so much so that organisation theory itself became part of the grand narrative of progress. If it is accurate to claim that there was a time in which the narrative of progress was dominant, and if it is correct to state that old modern organisation was part of the narrative of progress, then the emergence of any new narratives requires careful consideration, since they might challenge the foundations of modern organisational theory. Narrative analysis provides us with the empirical tools to achieve this. In the next section we define narrative analysis and explain how it can be used to identify and examine new narratives.

Narrative Analysis Defined

The relatively recent attention that narrative analysis has received as part of the emerging field of organisational discourse cannot conceal that it is in fact a very old and well-established mode of conducting research. Flyvbjerg (1998) points to Aristotle on the structure of stories and Callinicos (1995) and Martin (1986) refer to Herodotus's Histories as famous examples of narrative analysis that serve as studies of social reality. Nonetheless, and despite the fact that narrative analysis has been employed since time immemorial, it is not always properly understood and raises a number of questions. How do authors compose narratives out of shreds of stories? How do we know that a story is present, or even dominant, in a certain field?

Put in the simplest terms, a narrative is a story. However, if we accept that "stories have a clear chronological structure, with a beginning and an end", we find this structure is not always straightforwardly present in the narratives which we encounter in organisations. Instead, and in our experience, a more appropriate definition is that of the narrative as a "representation of a series of events meaningfully connected in a temporal and causal way", whereby an event is defined as the transition from one state to another state.

A further consideration when applying narrative analysis to the study of organisations is that it borrows concepts and tools from narratology—the literary discipline that theorises narratives and narrating. These concepts do not sit comfortably with the study of narratives in organising. The most important difference is that narratology studies text, whereas organisational discourse requires those applying narrative analysis to organisational phenomena to take a discourse analytic approach and therefore to consider the relationship between text and context. Furthermore, where narratology is particularly interested in the uniqueness of a story, discourse analysis in organisation studies takes a particular interest in alternative stories.

Policy Narrative

This chapter focuses on narratives which are used to underpin, defend, or attack certain policies. They meet all the requirements as stated in the general definition of narratives, but they form a specific variant—the kind of narratives which Roe (1994) calls policy narratives.

As such, they "are taken by one or more parties to the controversy as underwriting and stabilising the assumptions for policymaking in the face of the issue's uncertainty, complexity or polarisation". Whereas narratives can sometimes feature normative elements, policy narratives always feature a normative element. After all, policy narratives are not just accounts about a changing world. Instead, they prescribe the direction in which the world should change. They incorporate both a description of a problem situation and the direction of the solution. Schon and Rein (1994) have called this connection between a representation of reality and its consequences a "normative leap".

Policy narratives present arguments as to why a certain situation is problematic and how this problem situation should be dealt with. As with any narrative, this still involves a structured connection of events, but these events lie in the future. This process constitutes what can be termed the line of reasoning of a policy narrative. A policy narrative is also characterised by a recurrent issue that is continually repeated, developed, or referenced throughout the course of a narrative. This is the theme of a narrative. Examples in novels are "love" and "death," but in policy narratives themes such as "the public-private divide," "efficiency versus inefficiency," or "centralisation versus decentralisation" are likely to emerge.

A theme has the capacity to unify, though it can manifest itself in various patterns. Themes can consist of various elements, but by definition they have no plot. Whereas stories can narrate the relation between two unique events without reference to a recurrent issue, themes are always present in narratives. Authors have used different concepts to indicate the recurrent issue, for example "story line" and "motive".

The (re)Construction of Dominant Policy Narratives

One assumption underlying our research is that narrators—the agents who propose or attack a policy—do not start from scratch. They borrow a stock narrative and apply it to the particular connection between events. For instance, and in the context of the study in this chapter, a narrator located in the English water management industry told a story about a particular site in England.

He argued that in order to minimise the risk of flooding, a number of local rivers should not be straightened and dikes should no longer be heightened. Instead, bends should be brought back into the rivers and room should be made for the overflow of rivers using what are called retention basins.

This is a particular story, applying to one particular situation. At the same time, however, we have seen similar stories, told by different narrators, appearing on the stage with regard to other towns. Moreover, we have seen the story emerging not only with regard to rivers but also with regard to coastal defence in England and Wales. If these little stories, or petits récits, share the same theme (resilience), the same relation between events, and the same line of reasoning (e.g., "sustainable solutions are only to be found if you act according to the natural course, instead of acting against nature"), they are part of a wider, and sometimes dominant, policy narrative.

How do we know if such a narrative is dominant? This question points to the fundamental issue of the relationship between text and context, that is, between narratives, on the one hand, and practices, institutions, and nonnarrative accounts, on the other. To theorise this connection, we make use of Hajer's concept of "institutionalisation"—the theoretical concepts of a narrative are translated into concrete policies and institutional arrangements. In order to be dominant and finally hegemonic, a narrative needs to be institutionalised.

Water Management in England and Wales

Water management is concerned with tasks such as land drainage and flood prevention; water supply and sewage; pollution prevention; and river management. It is highly relevant to our analysis because it has historically been equated with the belief in progress. This chapter investigates contemporary water management in England and Wales. Studies in other countries have shown that the construction of the infrastructure for roads and water has been a prime example of modernisation conceived as the elaboration of policy narratives based on ideas of progress. If policy narratives other than progress are dominant in this field, it is an indication that other sectors which have not been put on a par with progress might also experience the emergence of new narratives.

Water services were completely privatized in England and Wales in 1989. Since then, drinking water has been supplied and treated by private companies and transported through and treated within a privatized infrastructure. The Department of the Environment, Trade, and the Regions (DETR) has general responsibility for the legal structure of the water industry and how it is regulated. The Ministry of Agriculture, Fisheries, and Food (MAFF) is responsible for land drainage and flood control. The services are regulated along three dimensions: economic, environmental, and quality regulation.

The Drinking Water Inspectorate (DWI) is the executive body for quality regulation. It sets standards of quality, both for drinking water and waste water. The Environment Agency is responsible for protecting and improving the environment. Economic regulation is the task of the Office for Water

Services (Ofwat). Ofwat sets the price on the basis of information, collected annually, relating to the activities of each of the water and sewerage companies. Water companies have to operate within the price limits set by Ofwat. The profit made within the price limit is an accepted profit for the water company. Ofwat has ten regional customer service committees, which are nonstatutory bodies that have been established to investigate customer complaints and to approach the companies and the director of Ofwat on matters which affect customer interests.

One could argue that there is a fourth regulator, the Monopolies and Mergers Commission. Any proposed mergers of water companies above a relatively small size are automatically referred to the Monopolies and Mergers Commission. Some cases have been blocked on the grounds that an adequate remedy for the loss of competitors cannot be found. In England and Wales there are ten water companies which provide both water and sewerage services and sixteen companies that supply water only.

Household consumers do not have freedom of choice concerning which water company supplies them water. As well as being answerable to the aforementioned UK regulatory bodies, water management companies in England and Wales also come under the auspices of the European Union (EU). European legislation plays a role in the setting of standards (drinking water supply, bathing water, and sewerage) and in the designation of nature reserves. Each member nation of the EU has to ensure that its water industry operates according to the laws described in the relevant European directives. A further important EU influence comes from its regulations designed to ensure open competition in various industries, sectors, and markets. The European Commission has decided that drinking-water supply should be subject to these regulations. Finally, there are a variety of environmental pressure groups with interests in the way water is managed in England and Wales. These include Friends of the Earth, the Royal Society of the Protection of Birds, and Surfers Against Sewage.

The Narrative Repertoire

This section describes the narrative repertoire of contemporary water management. Can we find the traits of the narrative of progress, namely rationality and control? Or are there other narratives central to contemporary water management? In 1999, we conducted what Czarniawska-Joerges (1997) has called "narrative interviews" with thirteen individuals of key organisations in English and Welsh water management. Most importantly, during this type of interview "both the structure (the plot) and the main concepts (metaphors) are chosen by the interlocutor rather than the researcher". The interviewees were chosen either for their key role in policymaking, or for their role in the public debate.

Our theoretical framework distinguished between petits recits and policy narratives. The interviews and other documents from the field are considered to be little stories. In these separate stories, we have searched for shared accounts. Policy narratives are constructed out of the little narratives according to the elements mentioned above: theme, relation between events, and the same line of reasoning. If little stories share these elements, they are considered to be part of the same policy narrative. Judged by the degree of institutionalisation of the stories, our research identified four dominant narratives. These are the narratives of "Comprehensive water management," the "Eurostory," "Water as God-given right," and "Water as commodity." In the remainder of this section we analyse each of these narratives in turn.

Comprehensive Water Management

The basic assumption of comprehensive water management is that all aspects of the use of water, and all users of water, are linked with each other and should therefore be managed integrally. The narrators of this narrative are environmental groups, the DETR, the Labour government, and the Environment Agency. In this narrative, water is viewed as a system in which all different aspects are intimately connected with one another. Rain, sewerage, drainage, drinking water supply, groundwater abstractions, irrigation, and cleaning up the beaches must, it argues, be looked at integrally. Since all aspects of water are linked, changes with regard to one aspect will inevitably affects other aspects.

Figure: *Designing and installing a comprehensive water harvesting system*

This will be the case particularly if one part of the system is isolated and dealt with in a way which causes an intense disturbance to the natural environment. For instance, if the length of a river is shortened by taking out the bends, which is easier for shipping, the risk of flooding may increase. Therefore, according to this story, one cannot adapt technical solutions for an isolated aspect within the water system. Because of the interconnectedness, tight human control as a means of achieving sustainable use of water loses its prominence. Furthermore, there is an understanding that we cannot fight against nature, because in the end we will lose the battle. There are numerous examples of the institutionalisation of this narrative. The first is the UK Water Summit in 1997, organised by the then-newly elected Labour government.

The outcome of this summit was a Ten Point Plan, which lists the activities needed to achieve sustainable use of water. A further example is "Taking Water Responsibly", a report that reviews the extraction of water from rivers, lakes, and other bodies and how the process is licensed. From 1963 onward, abstraction licenses have not had a time limit. "Taking Water Responsibly" proposes to make all licenses limited in time and reviewed on a regular basis. Thus, instead of taking its departure from "water owners," the current policy is about "sustainable use." A final example of institutionalisation of this narrative is that, unlike almost all other European countries, the Environment Agency in England and Wales is divided into areas of responsibility according to the catchment areas of rivers instead of the administrative units of local authorities. Different aspects of water are intimately linked and should therefore be managed integrally at the water system level instead of managed according to administrative boundaries.

A quote from an interview with a civil servant at the MAFF illustrates the comprehensive approach: Internal drainage boards want to get rid of the water. The internal drainage boards are there to drain. And that is what the drains do. If you don't have to worry about the consequences, if you are only talking about five farmers who farm that little valley, and their only worry is getting rid of the water, yes, then it is a solution. The ideal solution. But it may not be very helpful for Northampton downriver, where it is going to get flooded. So draining is not necessarily an holistic answer. But certainly within MAFF, we understand that just draining, without looking integrally, is not the best approach.

The Comprehensive narrative has evolved around the themes of sustainability and uncertainty. If all aspects of water are related to each other, a disruption in one area will make itself felt in others. Similarly, if there is uncertainty in one policy area, policymakers will have to deal with uncertainty in other policy areas. There are also many references in this Comprehensive narrative to the unpredictable effects of climate change. Further, there are questions about human behaviour, as a member of a customer service committee describes:

In this particular region, the demographic seems to be changing. We've got more single person households. Therefore you do not have the economies of scale. And the rainfall patterns—you just don't know. So from the water supply side of view, there are three areas of concern. One is the demographic changes, secondly global warming, and in the third place changing activities in the house. This brings about a lot of uncertainty. Within this narrative, we can find a specific story on the limits of water. It is a story emphasizing that water is a limited and scarce resource which should be used very carefully. According to this story, money will not buy us water, and building reservoirs is not the solution for the rising demand for water. The solution is a more efficient use of water. Demand has to be systematically lowered. Although the narrative has been around for some time, arguably since the oil crisis in 1973, it was a drought in 1995 which heightened public awareness of it. The drought almost led to a failure of a few water companies, notably Yorkshire Water, to supply their customers with water. Huge lorries transported water over the roads where delivery via pipeline had become impossible. Water scarcity consistently made the media headlines. The Labour Party, then in opposition, recognised that there was political capital to be made out of the issue and attacked the Conservative government about it.

This story goes on to tell us that we have to use water more efficiently precisely because water is a limited resource. An interviewee of an environmental group described it thus: We take the perspective that because there are problems with over-abstraction of water, too much water is being used. Therefore we have to reduce the amount of water we use in order to live more happily within our bounded resource. So our policy should be based around reaching that target.

These examples show that the Comprehensive narrative is institutionalised in contemporary water management. As such, it is one of the dominant narratives in the industry. It is, however, a nonmodern narrative. It contradicts the narrative of progress in several ways, not least because it is epitomized by uncertainty and irrationality. Water is becoming increasingly scarce and there is a need to lower demand, despite the fact that customers want more of it. The water system is seen as complicated, fragile, and integrated, where a disruption in one area can lead to problems elsewhere. Moreover, the vagaries of nature make its management by science and engineering particularly difficult. Given these stories, the narrative of progress is unable to offer an appropriate solution to these problems. The solution instead is to be found in a new narrative. We are told that a new, more coordinated approach to the management of water is emerging—one that views the system as a whole and attempts to provide comprehensive rather than isolated planning and solutions to its problems.

Eurostories

Eurostories describe the impact of European legislation on water management. The story starts with the investments in the water infrastructure that the UK government and water industry had to make in order to comply with European directives that came into force in 1989. Tremendous effort was needed in order to comply with the stringent European standards, particularly in relation to "water quality." Unlike the situation before 1989, when there had been the requirement to deliver "wholesome water," for the first time there were now clearly defined, minimal, scientifically calculated standards to meet.

The Eurostories are mainly told by water companies, and to a certain extent the Drinking Water Inspectorate, the Environment Agency, and the DETR. The stories have two themes. The first concerns the importance of central legislation for all member states. The second centres on comparison between countries. These themes are connected. Because of the central legislation, which prescribes similar standards and processes, achievements and results of the countries can be compared.

In relation to the central legislation, the narrative is full of complaints about the impracticalities of certain procedures of the European directives and the need to improve them. The Drinking Water Inspectorate and the umbrella organisation of the water companies in the United Kingdom consistently argued that the EU imposes new regulations without involving them enough. Early directives turned out to be highly impractical. They have continued though, and more recent directives such as the Framework Directive (overarching legislation applying to several aspects of water management) are believed to be much more flexible and practical than their predecessors. An admission of these early problems can be seen as emanating from the EU itself. The new Drinking Water Directive, which came into force in December 1998, specifically acknowledges that some solutions were impractical in the former version. According to an interviewee at the DWI:

The new directive...is now very much more sensible. There was the implication in the original directive that if you did not meet the standards, you had to cease supply. But ceasing supplies would of course have a much bigger impact on public health, then marginally exceeding a standard, which has got a huge factor of safety anyway. We have now got in this new directive the derogation position where you can effectively have a period to solve the problem. You cannot do it immediately, but you get a sensible period.

The tendency toward increased practicability is present in all comments on European legislation. In the process of making the Framework Directive for example, European legislation no longer tries to take care of all the details. Instead, it sets up a framework, and leaves it to each member state to fill in the details relevant to their own localised situations. The reason they turned to this approach is that the predecessor of the Framework Directive failed after member states had gone through sixteen drafts: "It wanted to achieve a state of perfection that could never be reached," according to an interviewee at a water company. The result, according to the water companies, the Environment Agency, the DETR, and the DWI, is that the Commission has now adopted a style of working which is far more effective. For example, an interviewee at the DETR stated: The way forward of course is the Framework Directive. This is not necessarily the formal government position, but ... it seemed like a good way forward in providing a framework, and only a framework, so leaving plenty of scope for the individual states to apply it to their own domestic circumstances.... The recognition that setting a broad framework, concentrating on those things which clearly are important on a Europe-wide basis, and leaving as much as possible of the detail to member states' discretion, is the way forward.

Investments enforced by European legislation, and the impracticability of certain directives lead to the next event in the narrative of Europe as the new legislator. This is about the necessity of influencing the legislative process in Europe and being present at the decision-making in Brussels. All parties ensure that their interests are represented at the European level, apart from environmental organisations which do not have the resources. As an interviewee at one such organisation explained: (The European directives) were certainly a big issue with respect to the pollution standards, pesticides and nitrates, and so on.... I mean, I was alert that their Fresh Water Directive came forward, but we just did not have the capacity to deal with it. So we have left it out really.

In contrast, and not surprisingly, the water companies have pressed very strongly at the European level, notably with regard to the new Drinking Water Directive and the Framework Directive. The water companies also have a European umbrella organisation, Eureau, which represents their interests by, for example, attempting to ensure that what comes out of the Framework Directive is practicable. The water companies in the UK have even reorganised their national umbrella organisation in order to respond to the EU bureaucratic structures. At one time there were separate bodies for Scotland, Northern Ireland, England, and Wales, but these organisations have merged into one. To use the words of an interviewee at a water company: "As far as Europe is concerned, the United Kingdom is just one country. So it makes much more sense just to have one organisation which is assisting the industry."

The European directives also have enabled a comparison between the member states. Comparisons are continually used in the discourse of water management. Now all member states use the European comparison to show their ranking in relation to each other. One of the problems with this is that England and Wales are more open in the provision of information that allows such judgments to be made compared to other countries. This means that they frequently receive censure and criticism as a result of providing information which other countries withhold. An interviewee at Ofwat made the following point about the allegedly high leakage of water from the mains system in the UK:

We have put quite a lot of effort into collecting the right level of information and understanding: what is leakage? We spent three full years in getting the data right. And then you get some very big numbers. People contrast that with "sure you cannot have 25 percent leakage, it doesn't happen anywhere else in the world." Our main concern is that the leakage in the UK is not worse than anywhere else.... You need to understand the numbers behind it. We have required companies to put in proper systems to do this. We have required them to challenge all their different assumptions. We required them to be publicly audited by technical experts. That is a lot different than some operational manager reporting to some civil servant stating it leakage is ten percent.

The Eurostory is shared by most parties within water management, apart from the environmental organisations. There are a number of further important observations to make about its status. First, European directives are imposed on the national water management of England and Wales. This forces organisations to focus on the European level with regard to lobbying, negotiation, image, and informal networks. If stakeholders do not organise themselves effectively at the European level, they run the risk that the European Commission will impose an impractical template on them without having consulted them.

Second, we have seen the European Commission act in the past as a central, rational body, imposing what it deemed to be best practice and the most appropriate regulations on all members. This approach seems to have rebounded on them and has led to the new Framework Directive—a directive that leaves space for variations in national approaches. The Framework is defined, but the details are to be filled in by the member states. There is agreement among the interviewees that this is an appropriate way forward. Third, all interviewees who used this narrative continuously compared the performance of other member states with what was occurring in England and Wales in terms of water quality, transparency of the process, billing procedures, price of drinking water, or the organisation of the sector. The theme of patriotism, so common to the old modem narrative of progress, is still present in the contemporary repertoire.

Finally, the narrative of the Eurostory can be viewed as a prolongation of the old modern narrative of progress, since it favours centralisation, Weberian like bureaucracy, and increased transparency as means to increase efficiency. It also has the important side effect of increasing patriotism, a specifically modern theme. On the other hand, one can discern ways in which it runs counter to the narrative of progress. Increasingly, templates and standardisation have been put aside, and there is a tendency toward more practical solutions that take into account local variations. In this respect, the new Framework Directive seeks to facilitate what is tantamount to the comprehensive water management system described above. In effect, elements of the Eurostory can be seen as nonmodern and as challenging rather than prolonging the narrative of progress.

Water as a God-given Right

The basis of this narrative is that water is the first necessity of life and should therefore not have been privatized as it was in 1989. Given that privatization has occurred, companies should not be allowed to make excessive profits at the expense of the captive consumers by making use of a resource that belongs to the community. Both customer service committees and environmental organisations use this argument, albeit for different purposes. Individuals who are not collectively organised also frequently express themselves—for example, in letters to the editor—using this narrative. There are two themes within this narrative: The first theme is that of the "fat cats."

This theme refers to the fact that companies make easy profits out of the provision of what is a public resource that they should not have been allowed to own in the first place. The second is the theme of public health. The provision of water is the first necessity of public health. Therefore, there should be no barrier to citizens using as much water as they need. Private companies set up such a barrier by imposing excessive charges for water based on "the market rate."

The narrative told by means of the fat cat theme is fairly simple: the government should not have sold any of the water companies, and where it did, it sold them far too cheaply. The shareholders and directors of the companies are now earning a lot of money without delivering a good service. Consumers in the postprivatization period have experienced increased rates; poor service, as evidenced by a high level of leakage; and huge unearned salary increases for the managing directors. No one should be able to own water, as we can see in a quote from a member of a customer service committee: "Here is a general resistance against metring... . There is this commitment to public service through welfare state. It would be wrong to see water as a saleable commodity."

The narrative of water as a right is often told by making use of the theme of public health. In the debate about the metring of water use, the DETR, water companies, and environmental groups have argued that people will use water more wisely if they have to pay for the amount they are using instead of a fixed sum based on the ratable value of the house. Thus, these parties consider metring to be a means of more fairly managing demand for water. However, some parties within water management, most notably the Labour government and Ofwat, think that metring is not always the best answer, because it has negative effects on the health of people, especially low-income members of society who have difficulty paying their bills and those with a medical condition that requires them to use large amounts of water. The new Water Industry Act (1999) safeguards rights of access to water.

Disconnection of bill defaulters is no longer possible, and metring has become optional instead of obligatory in order to protect the vulnerable customers. In certain respects, the act upholds many of the polices, practices, and values underpinning the management of water when it was a publicly owned utility. In sum, underlying the narrative of water as a God-given right is the perception that water is best managed and supplied where it is publicly owned. Only then will it be easily and cheaply available to those that already own it—the public. This view is taken irrespective of the likely bureaucracy, centralisation, and rationalisation—along with their attendant problems-that such an arrangement will lead to. In this respect it is a modernist view that is highly consistent with the narrative of progress.

Water as a Commodity

The narrative according to which water can be sold as a commodity is the counternarrative of the water as a God-given right narrative. Drinking water can be individually purchased and individually used. Therefore, it should be sold on the market. The most strident advocates of this narrative are the water companies, but Ofwat engages in this narrative as well. The commodity and water as a right narratives show an overlap in events. There are two accounts of the same process: one states that privatization has only been good for the managing directors' wallets, and the other states that both quality, the consumer, and the environment have benefited from this change.

The latter account asserts that privatizing public water authorities in 1989 was a good idea and turned out to be a huge success. Within a year, shares had generally doubled in value. It was easier than expected for the companies to raise money from the capital market, so capital investment was cheaper than expected. All in all, things went better than expected in the process of privatization. All of the attention in the public domain, however, was focused on the salaries of the top bosses of the water companies, on the problems with the drought, the leakage percentage, and rising bills. In defence of the water companies, the public did not realise that besides the decades of underinvestment in infrastructure, there was also a new European directive which required huge investments in the environment.

Moreover, consumers complained about the increase in the size of their bills since privatization, but they forget that their bills had already gone up steeply in the years before privatization. The narrative goes on to assert that with regard to price and quality; customers have had a good deal with privatization. Water quality has improved enormously. In 1989 English and

Welsh drinking water did not comply with all European standards. As we noted earlier, England and Wales were considered the "dirty men" of Europe. At present, England and Wales comply with most European and international water standards. Despite all these achievements, consumers do not believe that privatization has been a good deal, according to an interviewee at Ofwat:

The conundrum in the water industry here is that prices since 1989 have gone up, by about 30-odd percent. But what would have happened if we just stayed put, but also delivered all the European directive programs and improved services. What would that have cost the customers? The prices would have gone up not by 30 percent, but by 50 or 60 percent.... Investments essentially doubled. We are now compliant with most of the European directives.... For consumers, I argue, privatization has been a good thing. The difficulty is, they don't believe it.

In the early 1990s, any public sympathy toward the idea of water as a commodity was undermined by negative views on the privatization of water. At the end of the 1990s, a shift in repertoire occurred. Some interviewees argued that the election of a Labour government in 1998 played a significant role in effecting this shift. Criticism of the water sector had provided them with ammunition to attack the previous Conservative government while they were in opposition. Once they were in power themselves, they had to solve the problem instead of stirring up the debate. Now, they often emphasize the upside of privatization. Thus, the basic tenet of government policymakers in water management (primarily in Ofwat) is that privatization has worked out well, while even some environmental groups (e.g., English Nature) suggest that it has not worked out too badly. The tragedy, according to this story, is that the public does not believe it. The advocates of this narrative complain that newspapers are only interested in bad stories; they do not report the achievements.

If we compare the two counternarratives on water as a social service, "water as a God-given right" versus "water as commodity," we see an interesting difference emerging. We argued above that old modern organisation theory, which was itself part of the narrative of progress, held that efficiency was best achieved where the state acted as the central controlling and organising authority for society. Growing state influence is visible in water management in England and Wales during the period in which the narrative of progress was dominant. At present, the public nature of water management is hotly debated.

Although "water as a God-given right" continues the old modern narrative of progress as it supports the idea of the public ownership of water, "water as commodity" could also be seen as another narrative in the modern tradition with regard to its core theme, which is efficiency. However, under this narrative and in order to enhance efficiency of supply, water need no longer be provided by state -run organisations. Proponents of the narrative assert that a free market will safeguard efficiency. In effect then, the "water as a commodity" narrative contains an inherent contradiction that challenges modern organisational theory's perception of progress via public ownership. It does so by decoupling the efficient supply of water from the public/state domain.

Challenges to Modern Organisation Theory Posed by New Narratives

In the four dominant narratives that we have identified, we have looked for indications of the presence of the narrative of progress, with rationality and reason as central themes. We found optimism about human capability, but this was always undercut by the notions of risk, holism, and uncertainty. Even where the narrative of progress used to be dominant at the level of the European Commission, its influence has waned in recent years. Thus, we have no indication of institutionalisation of the narrative of progress. However, it should be remembered that while we found that progress is no longer dominant in the narratives of the present, it is not entirely absent either. It has been decentred rather than abolished.

The comprehensive water management approach and the Eurostory hold particular importance in the deconstruction of the underlying rationality of the organisation of water management. For example, it is noticeable that the new narrative of comprehensive water management brings uncertainty to the fore and makes water scarcity a key issue. These limits were not recognised at all in old modern theorising. Moreover, the comprehensive approach accepts that "nature" constrains technical possibilities. This places the experts of old modernity (scientists, engineers) in a difficult position, since it raises the question of whether water management is primarily a technical exercise. Those telling this story suggest that the narrative of progress is particularly susceptible to events—the drought of 1995 for example—which put limits on its agenda in a powerful way. Under water as comprehensive management, the appropriate response to all these problems is some kind of demand management which recognises the systemic links between rain, sewerage, drinking water supply, groundwater abstractions, irrigation, and the cleaning up of beaches.

The Eurostory highlights the investments needed for the UK water supply to meet the European directives that came into force in 1989. For the first time in the UK, there were "measurable standards" that needed to be complied with. The impact of these directives and the impracticality of some of the measures contained within them required lobbying and organisation in order to influence Brussels-based decision-making. The measurable standards and their achievement by the water authorities of member states allow for comparison across member states and designation of laggardly countries and water authorities. The comparisons encompass not just water quality but also billing procedures, pricing of drinking water, and the organisation of the sector.

Water as God-given right versus water as commodity form counternarratives with regard to water as a social service. The former is the most modern narrative to be found in the contemporary repertoire. It defends the public-private boundaries which have long been established by the narrative of progress. The attack of the old modern boundary is realised by the latter, which although it has efficiency as its core theme, suggests this can be achieved by minimising state influence over water management and by placing it in private hands. Our analysis of the four narratives of water management leads to the identification of several significant challenges for modern organisation theory. The first, and perhaps most fundamental, challenge stems from the very act of focusing on narratives at all.

Narratives demonstrate plurality, pluralism, and plurivocality. They encourage us to look for a variety of stories within organisations rather than a single monological account of what is going on, and as such, they challenge the modernist authority of any centralised singular conception of organisation. In short, the focus on narratives makes it much more difficult to believe that an organisation can be so easily gathered up into one voice, or captured in one formula. Second, the boundaries between the material and immaterial aspects of organising, which are taken for granted under modernist organisational theory, can no longer be maintained. The distinction between the two is blurred. In particular, the narratives show that the politics and process of organising and managing water are part of the same activity. What is therefore required is an "organisational political ecology" to be developed that respects the diverse materialities that impact on organising.

Third, the Eurostory narrative in particular suggests that the boundaries between the inside and outside of the organisation are increasingly hard to discern and either need to be disregarded or at the very least redrawn. For example, the internalities of the water companies now appear to include important externalities, such as the EU, which function as if they are internal parts of the organisation. This change in circumstances has challenged the narrative of progress, which originally dominated the Eurostory narrative, to the extent that the Eurostory now exhibits themes consistent with the new narrative of comprehensive water management.

Fourth, the certainty of modern organisation and its focus on rationality no longer holds. Organising is no longer a certain activity and the guardians of certainty (scientists and engineers), are no longer deemed to be the primary force for progress. Organising is no longer a technical exercise but in many ways an ongoing attempt to respond to a variety of factors such as "nature," governments, "customers," and European legislators. In sum, uncertainty bequeaths us a risk-filled society in which certainty for the organisation can never regain its centrality. The fifth challenge to modern organisation theory concerns the issue of delinkage. We have seen that the water as a God-given right narrative and its counternarrative, which considers water as a commodity, have dealt differently with privatization.

The water as a right narrative expresses a view that resources such as water merit being supplied by a public body and that to do so is in the "public good". Its proponents claim that this duty cannot be entrusted to a private organisation, which will misuse its monopolistic position. On the other hand, the counternarrative assumes that such resources can be provided by agents other than the state, in this case private companies. This is an example of what Albrow has called "delinkage".

Delinkage indicates the disruption of the links between previously accepted structures, principles, and practices. Rationality used to be the linking factor between these three elements. However, organisational structure no longer corresponds to the boundaries between different practices; basic principles do not provide for the best possible practice; and structure has freed itself from principle. Delinkage as exemplified by water management, therefore, challenges the basic tenets of modern organisation theory. New practices, principles, and structures have been invented in the organising of water management now that other agents perform duties which have been the responsibility of the nation state for the last hundred years. Modernity has become decentred.

The Response by Organisation Theory

The challenge of delinkage is perhaps the biggest challenge of all to old modern organisation theory. We have seen that structure, principles, and practices are constructed in the process of organising. These elements are no longer linked and do not form a template that can be applied to every circumstance. Instead, delinkage has occurred, whereby organising is shaped according to the suitability of situation and place. In the case of water management in England and Wales, delinkage became a key issue after privatization. New agencies have taken over tasks which for over a century were the exclusive domain of the nation state.

Organisation theory has responded to these challenges in complex ways. The flexibility, fluidity, and performativity of organising have been brought to the fore, as has the importance of organisational boundaries, time, and space. Organisations are plural artifacts which can no longer be accounted for by rationalist or merely internalist accounts, but need to be explained by a kind of complex hermeneutic ecology of organisation. Some scholars have looked for grand reversals, suggesting that where the modern organisation was built upon the division of labour and other key differentiations, it should now be analysed on the basis of de-differentiation. Many writers have enjoined the postmodern perspective which entails deconstruction, critiques and problematization of power, and giving voice to silence(s).

Others, in trying to provide the continuities required of textbook introductions to the subject, simply list all the scholars and perspectives that have influenced organisation theory from "classical" to "modern" to "symbolic-interpretive" to "postmodern". We propose an alternative way forward—one that is sensitive to the challenges we have identified and which incorporates narrative analysis. Dunford and Palmer (1998) bring out some of the paradoxes that arise when a narrative or discourse-based approach is used to understand organisations. They accept that not all of the paradoxes are actually paradoxes, nor do they all have the same status in relation to the problematics of contemporary organisational thought or practice. These paradoxes are that narratives can be different from action; that narratives both create and destroy meaning; that they are both specific (unique?) and general; that non-sense is also sense; that organisation is narrative; and finally that narrative both enables and constrains action.

Several of these "paradoxes" have as their basis the attempt or need to bring together agency and structure. Narrative analysis requires us to link narrative as action to narrative as structure. It also requires us to consider those older sociological questions such as whose motives and interests are being served by a narrative when it is mobilised; and to what effect and for what purposes. Narrative and the use of narrative-based approaches should not, however, be seen as providing some magic (re-)solution to all our questions. Some of the paradoxes will melt away in relation to this disciplined work, some will pragmatically be rejected as too trivial, others will remain transhistorically as reminders that some issues are intellectually "solved" only for the moment and do not accede to a general resolution.

If, as Clegg has argued, organisations were at the heart of the modern condition then fundamental changes in that condition registered by the transformation of dominant narratives, such as the role of uncertainty in the comprehensive water management narrative, mark the transformation of contemporary conditions. Contemporary organisational analysis must find the analytical and symbolic tools with which to examine and understand these changes. However, our reflections on the paradoxes of discourse and narrative-based approaches suggest that more is needed. We do not believe that we have changed from a modern to a postmodern condition but rather that we inhabit a period in which the suppositions of modernity are no longer centre stage and are dissolving-they have not disappeared. Thus, there are aspects of modern organisational analysis, such as coherence and completeness of analysis, which merit use.

Organisational analysis remains as difficult as ever, and to focus on narrative is not enough. More than that, we have to resolve dilemmas of narrative and action and structure and agency, as well as the many different levels, temporalities, and strategic applications in which they arise. What we can be assured of is that a disciplined narrative approach that still asks difficult analytical questions of both itself and the relation of narrative to structuration and the construction of the organisation outside of narrative can begin to be an important resolution to the complex pragmatic and theoretical circumstances of contemporary organisation. Accordingly, this chapter has analysed both the "text" of new narratives and the context in which they are produced and used in order to identify and establish the new links that now exist between the structure, principles, and practices of organising. In doing so, it has shown that such an approach is of considerable value in showing how relationships that were once taken for granted within organisational theory now need rearticulation, since they have in fact become indicative of a dissolving framework for organising.

The Prospects for the Beneficial Use Doctrine to Achieve Efficiency in Western Water Use

Stretching existing water supplies to meet growing and changing demands will be the central challenge of western water management in the new millennium. Barring meaningful reforms, the century-old beneficial use doctrine and its corollaries, waste and forfeiture, are not the right tools for the job. This trio of doctrines accomplished certain purposes that were important in the late 1800s and early 1900s, albeit with varying degrees of success, but the doctrines really were not designed to maximise efficiency in water use, and their record of performance toward that goal is poor indeed. How could beneficial use, waste, and forfeiture be reformed to better achieve efficiency? Just as implementation of the beneficial use doctrine has been divided among courts, legislatures, and agencies, each of these branches could play a role in tightening up the beneficial use requirement.

Judicial Agenda

The above concluded that the judicial role in applying, much less improving, the beneficial use doctrine has been limited and constrained. Courts can do no more than decide the cases brought to them. Even taking all of the western states together, it is unlikely that many of the individual cases filed in the next few years will require the interpretation and application of the beneficial use, waste, or forfeiture doctrines. Nevertheless, because of the number of general stream adjudications currently working their way through the courts of various western states, the courts do in fact have an important role to play. In Arizona, two adjudications currently underway encompass two-thirds of the state, Idaho is currently adjudicating a single river and groundwater basin that covers ninety percent of the state. Montana is adjudicating practically all of that state's water. New Mexico and Utah are each conducting adjudications that encompass thirteen different rivers; New Mexico's cases affect major cities, towns, industries, and numerous Indian and Hispanic communities Washington and Wyoming are engaged in ongoing statewide adjudications, and most other western states are in the midst of at least one significant adjudication.

This heavy schedule of ongoing general stream adjudications means that the courts of several of the western states have before them, or will at some point in the near future, large numbers of individual water rights claimants. Without exception, it is the obligation of a court in a general stream adjudication to decree the appropriate amount of each and every claimant's water right. Every state has the same general requirement of beneficial use without waste as the basis, measure, and limit of every water right. Thus, it is incumbent upon the courts in these cases to consider the parameters of the beneficial use doctrine and prohibit waste in the water use practices before them. Of course, the degree of scrutiny that any particular court gives to existing rights and levels of efficiency will depend to a great degree on what approach the state water management agency takes to the issue, as well as on the arguments of the nonstate parties before them.

Nevertheless, a court has ample power and authority to request briefing and evidence on these issues if necessary to aid in its decision, regardless of whether the parties advocate active examination of existing efficiencies in water use. As a practical matter, these adjudications are all at different stages of the proceedings. Many of them may be well beyond the stage at which issues of efficiency can be raised and fully developed. But to the extent that the issues are still relevant and timely, judicial scrutiny is certainly appropriate.

Throughout the past century, the courts have been instrumental in the growth and adaptability of the beneficial use doctrine, to whatever incremental extent that has occurred. New uses have been recognised as beneficial to reflect changing social values and broadening scientific understanding.

Particularly wasteful practices have been curtailed, with a nudge to the water users to become consistent with changing practices around them. It is completely appropriate for a court reviewing water-losing practices, even those that may be customary, to question whether the customs themselves are wasteful. This is simply a matter of asking the next question, instead of taking for granted that if everyone is doing it, it must be okay. The courts in Imperial Irrigation District and Grimes did no more than this. Exhorting the courts to scrutinize water rights for waste is hardly asking for judicial activism. Instead, by doing so, the courts would be recognising the beneficial use doctrine for what it is—an element of the common law.

The common law is touted to be flexible and adaptable, and though accountable to precedent, not a mummified relic. What constitutes waste in 1999, with the western population about to pass an estimated eighty-eight million, ought to be both quantitatively and qualitatively different than what was recognised as waste in 1900, when most of the nation lived in the East. Further, it is not unreasonable to expect the perception of waste to be considerably different at a time when sophisticated water saving technologies are widely available compared to a time when primitive devices represented not only the norm, but the only choice. Also, eminently reasonable is the expectation that water use practices should adapt to advances in scientific knowledge about the importance of water in situ.

Many courts might agree with the general principle that the beneficial use doctrine should continue to evolve and not be used as a shield for inefficient water use methodology, no matter how widely practiced. However, they might balk at the next step of translating that general principle into a specific ruling declaring that any particular user's operation is losing too much water.

A court might feel that in order to make such a ruling, it would need to tell the user what level of loss would be acceptable, which would require the court to become expert in whatever the water user's business was and then tell the user how to run the operation. But a court does not need to go that far. It is enough simply to find that the current practice is wasteful, and if it is customary, that the custom itself has become wasteful. It is then entirely up to the water users (perhaps in conjunction with the state water agency) to determine what changes in method, technology, or practice would reduce the losses.

Another aspect of some of the recent waste cases deserves emphasis as an area for courts to consider in their review of water rights, especially in general stream adjudications. In the Alpine case discussed above, the court said that a water use should not be "'unreasonable' considering alternative uses of the water." This way of looking at whether a use should be considered beneficial has not received much discussion, but it represents a reasonable extension of the inquiry. For example, take a water user similar to Mr. Anderson in the SRBA case discussed above. The farmer had been using the same methods of "border" irrigation for sixty years. Sixty years ago, the primary alternative use for the water would probably have been another irrigated farm. In 1999 however, alternative uses include unmet senior tribal needs, as well as the dire need for increased instream flows to support endangered fish species, to provide adequate wildlife habitat, to manage water quality problems, and to satisfy recreational and aesthetic demands, before even considering additional consumptive demands. It seems that a judge could certainly defend a decision which asked the simple question: reasonable compared to what? and concluded that, in the current context, historically acceptable—even customary—flood irrigation practices are no longer reasonable and beneficial. Nearly twenty years ago, one commentator suggested a helpful five-step waste analysis for a court to employ in the irrigation context:

1. The protectable water right is the right to accrue benefits from watering crops.
2. Once technologies have developed that allow more efficient conveyance and application of water, the use of old, water-intensive practices becomes a privilege rather than part of the vested right.
3. With full appropriation of local supplies, the privilege is lost and use of the excess water becomes waste.
4. Once the statutory forfeiture period has run, the wasted portion reverts to the state.
5. Courts (and agencies) should determine how much has been forfeited as waste and allow diversion of only as much water as is reasonably needed under modern practices.

This suggested line of reasoning remains commendable and valid today and has application to all water users, not just irrigators. Although the courts are not the likely source of a revolution in western water use efficiency, it is perfectly reasonable to encourage the western judiciary, especially those courts involved in general stream adjudications, to continue the evolutionary process of adapting the beneficial use and waste doctrine to life in the twenty-first century. Suggested roles for administrative agencies and state legislatures are outlined in the next two sections.

The Management of International Waters: A Global Policy

In recent years an emergent global policy discourse has promoted the concept of Integrated Water Resources Management (IWRM) as a strategy for the sustainable management of international waters. However, integration remains a considerable challenge in large international river basins. This chapter addresses the relevance of the global discourse on IWRM, asking how much integration can be expected and how much integration is desirable in international water management. The chapter presents a novel compilation of eighty-six international river basin organisations and examines their degree of integration in terms of three dimensions: membership, substantive scope, and form.

More particularly, the chapter examines the integration problem from an economic perspective, asking whether integration serves the self-interest of the respective riparian states. The empirical evidence highlights the difficulties of integration, as the majority of international river basin organisations remain narrow in membership and scope. Economic considerations suggest that voluntary cooperation in river basins is institutionally demanding and that the degree of integration depends on the problem at hand. Hence, the challenge for international waters management is to search for the economically desirable degree of integration in each case.

The Problem

Some 263 river basins on earth cross international boundaries. Whenever a water resource is international, competing claims over shared waters can generate conflict among riparian countries. In response, a global policy discourse has emerged that promotes Integrated Water Resources Management (IWRM) as a concept for sustainable resource use. IWRM calls for the integrated management of different water uses within the hydrological confines of a river basin. Proponents of IWRM often advocate the establishment of special river basin organisations. However, a number of social scientists have rejected IWRM as unrealistic with respect to international waters. For example, Frank Marty, basing his predictions on experiences with six bilateral agreements, expects that the majority of international water agreements will be bilateral and narrow in scope and that this is indeed how it should be. John Waterbury warns that the quest for integrated development in international river basins will be elusive and excessively costly, although he concedes that bilateral agreements are not necessarily a solution for multiparty basins. Empirical research suggests that existing arrangements are limited in membership and substantive scope, but large-N analyses are rare and generally limited to the analysis of membership. In terms of membership, Jesse Hamner and Aaron Wolf find that 86 percent of 145 international water treaties concluded since 1945 are bilateral; however, they do not distinguish between bilateral and multilateral basins. According to Ken Conca, Fengshi Wu, and Ciqi Mei, thirty-three of forty-nine international river agreements established in multilateral basins between 1980 and 2000 are bilateral.

In terms of substantive scope, Ludwik Teclaff suggests that the majority of international water institutions are single-purpose, but the basis of his analysis is unclear. In terms of form, it is an open question as to what extent riparian states not only conclude agreements but also set up specific river basin organisations. Hence, only vague knowledge is available concerning integration of existing international river basin institutions in terms of substantive scope and form. How much integration can be expected in the management of international waters? How much integration is desirable in terms of membership, substantive scope, and form? In other words, how viable is the IWRM concept for the management of international rivers? To address these questions, I discuss in this chapter IWRM in the light of empirical evidence and economic considerations. I begin by setting out the global policy discourse of IWRM and then present the findings of a review of 506 international water treaties and eighty-six associated organisations.

This survey provides a more comprehensive empirical review of the degree of integration in existing international water management institutions in terms of different dimensions of integration, including membership, substantive scope, and form. I next examine economic perspectives on the conditions for voluntary integration within the international system, asking whether integration can be expected to be in the self-interest of riparian states. In the concluding section, I discuss the findings and highlight research and policy implications. The empirical review discussion confirms that the majority of international river basin institutions are limited in membership and scope, but it also suggests that this majority is smaller than is generally assumed in the literature. The economic analysis suggests that full cooperation in fact cannot necessarily be expected to occur. I conclude that full integration across international river basins is not necessarily desirable but that the challenge is to find the economically desirable degree of integration in any given case.

Global Policy Discourse on Integrated Water Resources Management

Starting with the 1992 International Conference on Water and the Environment, in Dublin, IWRM has been promoted at international environmental and freshwater conferences in Rio de Janeiro (1992), The Hague (2000), Bonn (2001), Johannesburg (2002), Kyoto (2003), and Mexico (2006). IWRM is the central idea of a global policy discourse supported not only by international organisations such as the United Nations and the World Bank, but also by new global actors such as the World Water Council (WWC) and the Global Water Partnership (GWP). The IWRM discourse is what other contributors to this special issue call a "mobius-web style" of global governance. IWRM has evolved as a response to traditional approaches to water resources management, which relied primarily on narrow engineering and sectoral solutions.

The discourse calls tegrated approaches, both in respect of the resource and among the actors involved. In general, there are different understandings of what is meant by IWRM. A very broad definition is provided by the GWP, which defines IWRM as "a process which promotes the co-ordinated development and management of water, land and related resources, in order to maximise the resultant economic and social welfare in an equitable manner without compromising the sustainability of vital ecosystems." Often IWRM incorporates the concept of river basin management (RBM), namely, that water resources should be managed in the context of a river basin. RBM is an institutional response to problems of "spatial fit" between hydrological and political boundaries. RBM suggests that water should not be managed at the level of geopolitical entities but at the river basin scale. In the case of international waters, RBM implies that all riparian countries should work together to manage water resources at the basin scale.

Although the concept of RBM has a longer history, it has gained fresh momentum over the last few decades. For instance, since 2000 RBM is a legal requirement of the European Union's Water Framework Directive (WFD). The concept has been supported by natural scientists, lawyers, and often also by economists. For example, Peter Rogers states: The general economic prescription to deal with externalities is to "internalise" them. The river basin itself is an ideal unit of analysis to achieve this goal: it can reasonably be assumed most externalities are captured by analysing the river basin as a single unit. This is why the concepts of integrated river basin planning and the creation of river basin commissions to implement and plan are so popular in the economic and planning literature.

A second interpretation of IWRM places greater emphasis on the integration of different water-using sectors, such as water supply and sanitation, industry, agriculture, nature protection, energy, and transport. Water is a multifunctional resource. Competing claims of different water users and water-using sectors create problems of horizontal "institutional interplay". In this case the challenge is to coordinate between different water-using sectors. To solve problems of fit and problems of interplay, there is often an explicit or implicit assumption that it is desirable to establish new agencies, so-called river basin organisations (RBOs), to institutionalise a river basin management approach. In principle, RBM could also be achieved through other means of coordination, such as freestanding negotiations. This notwithstanding, influential bodies such as the World Commission for Water for the 21st Century clearly advocate RBOs: "If the IWRM principle is adopted, then basin-level systemic management is clearly needed. ... Accordingly, governments should set up management agencies at the basin and aquifer levels."

Hence, IWRM encompasses three dimensions of integration: that of jurisdictions; that of sectors; and that of institutional arrangements. In the case of international rivers, the discourse on IWRM correspondingly points toward three different dimensions of institutional design—namely, membership, substantive scope, and form. Membership is constituted by those riparian states sharing a river basin that sign an agreement. Substantive scope refers to the range of water uses covered by an agreement. Form refers principally to the question of whether riparian states rely on agreements alone or whether they also establish corresponding organisational structures and international secretariats (sometimes also in collaboration with nonstate actors). In this chapter, the term institution is used as an overarching term referring to the "rules of the game." These rules are often manifested in agreements or contracts and include the constitutional documents of formal organisations.

IWRM has been recommended for international as well as domestic river basins. However, especially in large international river basins, integration tends to entail considerable complexities. In the case of international waters, "integration" cannot be imposed from above but relies on voluntary negotiations among the riparian states (and possibly other actors) that share a river basin. This raises the question whether it is in the interests of the respective riparian states to integrate. Therefore, the following empirical findings and economic analysis will assess the incentives for riparian states to integrate.

9

Rare Resources and Crises

The first area selected for ethnographic enquiry is the southern region of the Dra Valley, which corresponds to the last two oases of Ktawoua and Mhamid. This area represents the proper Saharan part of the Dra Valley and is most affected by drought and desertification. At the same time, it is the most isolated and, despite real needs, also the least studied and less impacted by 'development projects' than the other four oases of Mezguita, Tinzouline, Ternata and Fezouata. The study focuses on the access to and management of natural resources—mainly the rarest one, water—in relation to the social and economic features of the communities living in the area.

It is also concerned with the interactions between nomad and sedentary peoples, both in the past and today, the process of sedentarisation among nomads; the transformation of the pastoral economy, and the differences and similarities between the strategies used by nomadic and sedentary domestic units for coping with crises. After a broad survey of the wider region, stretching from the southern High Atlas to the Saharan borders in October 2000, fieldwork among the Ait Unzar was carried out between February and April 2001 and again, between September and December 2001.

The Context of Research: South-Eastern Morocco and the Dra Valley

Three main forms of economy have existed in the Dra Valley since early times: agriculture, animal husbandry and trade. For ecological reasons, agriculture has been restricted to six oases that benefit from the hydrological resources of the valley, especially the periodical floods of the Wadi Dra. The date palm is the most productive and widely cultivated plant and has the highest market value, but other annual crops—cereals, lucerne, legumes, vegetables, fruit trees (in the north)—have always been cultivated, mainly for domestic consumption. Until recently, the most common form of pastoralism was extensive herding. Its economic and social framework was well adapted to the extreme ecological environments of desert lands and mountains, and to the consequent scarcity and irregularity of rains. Various nomadic populations occupied the huge grazing lands surrounding the six oases of Wadi Dra. Intensive herding also existed, and was practised both by peasants in the oases and by settled nomads. Finally, the valley played a major role in the trans-Saharan trade between western Africa and the Maghreb, the Dra being the caravan route leading from Timbuktu to the northern cities and coasts of Morocco.

These ecological and economic configurations affected the political relations between people of the valley. Various dynamics of complementarity, conflict and subordination between: sedentary peoples and nomadic groups; different nomadic communities; and people of the Dra and the Moroccan ruling dynastics or centralised state authorities, marked the history of this area. The hegemony of Arab Bedouin groups coming from North Africa was well consolidated in the region by the medieval period, but in the seventeenth century the southern Berbers, previously dispersed by the Maqil Arabs, gathered in the huge Ait Atta confederation and again played a dominant role from the High Atlas down to the Dra Valley. Since these times 'protection' pacts (raya) established between the farmers living in the qsar (fortified villages) of the oases and the nomads became common.

The Dra region has also been characterised by a relative autonomy from the central power of the Makhzen, so much so that it has been called the bled assiba, meaning 'the country of insubordination'. This characterisation persisted during French colonisation. The Ai't Atta especially were among the toughest to resist the French occupation, and the 'pacification' of southern Morocco was achieved only in 1936. Although the French colonial administration tried to discourage, and even abolish the raya protection pacts, while pushing nomads towards sedentarisation, these institutions were respected until the 1970s, and even now they have not been abrogated completely. Where they have been abolished, their past importance has structured the economic, political and social relations between groups and continues to play an important role.

The characterisation of the Dra region as a crossroads points to a complex heritage of socio-cultural components. People are classified into different categories whose differentiation depends on various—sometimes overlapping—parameters: the most visible of these are the ethno-linguistic (Arab and Berber) and economic features (nomad pastoralists and settled farmers, or traders). However, 'status identities', according to which peoples are perceived and ascribed to a hierarchical framework, are the most important. The most common are the "abid, former slaves from sub-Saharan Africa, the Draoua or Haratin, sedentary farmers of the oases villages and the Chorfa and Mrabtin groups with religious functions. Obviously, identities are not static and the social relations attached to each of these can vary from one context to another and from one period to the next; but their long standing persistence, the strong relationship between the people of each category and their access to natural resources and power is completely accepted.

Modernisation during the last decades has brought about changes in the economic setting of the Dra Valley, even though they are not as fundamental as in other Moroccan regions. While the trans-Saharan trade lost its importance at the end of the nineteenth century, with the fall of Timbuktu and the beginning of colonisation, agriculture and herding remain the basic sources of income in a region that has not participated in the process of industrialisation. No doubt, roadbuilding, national and international migration flows, the processes of sedentarisation and urbanisation, as well as the development of tourism, are all factors that influence the socio-economic life of a region that has remained one of the most rural of Moroccan provinces." The building of the Mansour Adh-Dhabi dam (completed in 1973) which controls the stocking and distribution of water in the Dra Valley from Ouarzazate is perhaps the most important change in the life of the region: policies and decisions of the central state concerning water now take the place of climatic and other natural contingencies.

The Place of Ait Unzar in South-Eastern Morocco

Since the seventeenth century the Ait Atta confederation has been the dominant Berber group of the Dra valley south of the Atlas. Its history and political organisation are well-known due to the work of anthropologists in the 1960s. To recapitulate briefly, the Ait Atta confederation has a structure that is articulated in fifths (khoms); each fifth is further divided into various tribes (taqbilt) and these tribes into lineages (ikhs) of approximately five-six generational depth) Though in the last decades, after the struggle against colonialism, the military unity of the Ait Atta has no longer been required, this huge group has various features that are pertinent for community life, these include:

- the nomadic pastoral connotations still existing even for the more settled groups within the confederation, and thus implying the existence of a collectively appropriated grazing territory and the conclusion of annual agreements for the seasonal exploitation of the confederate pastures from the southern High Atlas to the Sahara;
- the existence of a common court functioning with a written customary law applicable to every Ait Atta tribesman;

- numerous social and cultural features that distinguish an attaoui from other southern Moroccans, sedentary or nomad, and, finally,
- a persistent trend towards tribal endogamy.

However, even with this cognitive unity and common identity, until recently harsh conflicts often took place between the different tribes of the Aft Atta in their struggle to ensure dominant economic and political roles. The Ait Unzar are part of the third Aft Atta khoms and are divided into two subgroups, the Aft Isfoul and the Ait Alouane, each half being further divided into four taqbilt. Thus, the Ait Alouane are further divided into the tribes Ait Ghenima, Ait Unzar, Ait Boumsaoud and Ait Sidi. The Ait Isfoul gained hegemony over the villages of the Ktawoua oasis during the second half of the nineteenth century; most of them are now settled in this area, some even since colonial times.

During the same period, the Aft Alouane established their power in the marginal zones of the southern part of the Ktawoua oasis and in the Mhamid oasis, from where they were partly evicted shortly before the French colonisation, by the 'Arib, an Arab nomad group that came from the Sahara. The Ait Unzar are still considered to be one of the 'most Saharan' groups of the Ait Atta confederation and this implies certain characteristics that tend to distinguish them from other tribesmen. First, there is the importance of camel herding to their pastoral system, rather than the goat herding that dominates among most other Ait Atta nomads. Second, is the extension of their territory and their migration cycles: unlike all other Aft Atta sub-divisions and tribes, who since historical times own rights to both the northern mountain pastures (in summer) and the southern plain pastures (in winter), the Ait Unzar do not possess specific tribal rights on the mountains pastures, and their cycle is more horizontal (with east-west movements in the southern desert area, according to rain availability) than vertical (with upwards movements exploiting the difference in altitude).

This also means that it is difficult to define their regular annual cycle of pastoral movements, which depends strictly on the Sahara's high unpredictable rainfall. This in turn, implies an annual variability of movements and a wider dispersion on the territory, in search of pasture. The third element that is closely linked to the ecological and economic characterisations mentioned earlier is their stronger socio-cultural proximity with southern Saharan nomad groups, such as the 'Arib. Berber identity, and within that the Ait Atta one, is very strong and often evoked, even though the distinction between the Ait Unzar and the Arab groups is stressed much less than among other Berber nomad groups. Even illiterate people know and speak the local Arabic dialect, as a second language after Tachelhit. Dress—another point of strong self-identification among the Air Atta people—is often the same as among the Arab

Saharan groups: the long, coloured malakhfa that women wrap around themselves and the blue khent tunic worn by men. While marriages take place among the Ait Unzar and within each similar Aft tribe, the rare extratribal unions are largely with Saharan Arabs.

Finally, the resistance of the Ait Unzar against complete sedentarisation must be noted. Although they are affected, as other nomads are, by the increasing trend towards sedentarisation since the mid 1970s; compared to other Berber nomads in the area, the Aft Unzar still have a fair number of families who mainly live by extensive herding and move with tents or, in some manner, maintain part of the group with their herds on desert pastures, while the other part is settled in 'modern' houses. It is indeed remarkable that in spite of the ideologically long lasting trend to erase the presence of nomads from statistical and other official documents, among the forty-two administrative divisions of the Commune Rurale de Ktawoua, the Ait Unzar are the only group considered and classified as 'nomadic' even in the last census of 1994.

A Case Study: The Ait Unzar Nomads and the Village of Tiraf

Even though the Ait Unzar were basically nomad pastoralists until at least the 1970s, today their socio-economic strategies range from long-distance mobility to complete sedentarisation. In my study I chose to examine different types of Ait Unzar domestic units, which could be considered representative of these various strategies. Without intending to establish a rigid taxonomy, one can broadly classify these households in the following categories:

1. groups living in black goat-hair tents whose subsistence still mainly derives from extensive herding and date gathering and who have a high annual and interannual mobility;
2. groups among whom the domestic unit is divided in two: one part taking care of the herds and following traditional nomadic management patterns, and the other part engaging in wage labour and settling in fixed dwellings of some kind;
3. groups who chose to settle in the 1960s and 1970s and for whom extensive herding is no longer the basic productive activity and spatial mobility of the family no longer a practised option; other sources of income and patterns of life have progressively become dominant.

The fieldwork focused on families who continue to be nomadic in the regions of Anagam, Tal Maidrin and Oued Mird, and some settled or semi-settled families in the urban centre of Tagounite and in the village of Tiraf.

As already mentioned, one of the major factors making it difficult to define a repeated cycle of pastoral movements among the nomadic Ait Unzar is the high temporal and spatial unpredictability of rains in the Sahara. Though a general grazing area was historically claimed and taken by the tribe, the movements within it are not regular. If the need arises, the Ait Unzar can also search for pasture outside this area and at the same time, other tribal groups are allowed to graze and access water in what is formally Aft Unzar territory.

The flexible character of Ait Unzar cycles has persisted since ancient times—and is even increasing with the recent droughts. The name of the group also seems to carry part of its history: the first term, Ait, is the plural of igwma, 'brother' (literally 'the son of my mother') currently used to define descent or tribal groups, while anzar (becoming unzar in the genitive construction) means 'rain'. So the terminological description of the group is something like 'people of the rain', and some Ait Unzar explain this by saying '... we follow the rains, deciding every time where to move with our herds; that is the reason why we are called Ait Unzar'.

In fact, in recent years, pasture availability has been one of the main constraints for moving: In Ait Unzar territory, permanent wells are not a major problem and in times of drought they can complain about the absence of rain water for pasture growth rather than about the lack of ground water. Though in the last decades the drought prevailing in the area seems to be the principal determinant of Ait Unzar seasonal strategies, nomadisation patterns are also highly conditioned by non-ecological factors.

Political and military measures are the most relevant for the limitation of pastoral space: the problems at the eastern border with Algeria since the 1960s and the conflict with the Polisario in the Sahara since 1975 brought remarkable change into the lives of the Ait Unzar and other Saharan pastoral groups. As of writing, the area considered as effective Ait Unzar territory is limited by the Jebel Bani to the north, the Oued Mird south of Takatt Ait Alouane to the east, the region of Mhamid to the south, and Foum Zguid to the west. These Ait Unzar nomads are scattered during most of the year on the grazing lands within this area, in autumn most of them approach the oasis of Ktawoua to participate in the date harvest.

The Ait Unzar also have a fixed place of attachment in the oasis, like most of the nomadic Berber groups of this region. It is the village of Tiraf, in the extreme south-eastern fringes of the Ktawoua. A pact dating back to 1856 established the exclusivity of the Ait Unzar as 'protectors' of the Draoua settled farmers who inhabited the village. In exchange for such military protection the farmers ceded them one part (normally one quarter) of their qsar dwellings and cultivated lands, including water rights for these. Even though these raya pacts have been formally abolished, the people of Tiraf, both nomads and farmers, consider their own pact as still valid. The present population of Tiraf still consists of the double component of Berber Ait Unzar nomads and Arab farmers of different Draoua families, and both groups refer to this original pact as the basis of their current cohabitation and overall social relations.

Even the local institutions illustrate the sort of 'bipolar system' of two qabila (tribal assemblies) who meet regularly to discuss and take decisions about the basic concerns of collective socio-economic life: agricultural season, sharecropping, contract between nomad guardians and farmers during date harvesting, water management, drinking water distribution, school services, etc. Therefore, within the village, the importance of long-standing nomad-sedentary relations remains visible in daily matters. For the Ait Unzar, since the pact of the mid-nineteenth century, Tiraf represents the pole of reference for the economic strategies of their community: most Ait Unzar own fields and houses in Tiraf, and can continue to practise extensive herding on the surrounding tribal pasture lands, while at the same time maintaining the opportunity of agricultural income, thanks to their ownership of cultivated plots in the village.

Here the geographical location of Tiraf is significant, as it is still within the oasis agricultural area of Ktawoua, but at its south-eastern fringe, projecting eastwards and southwards on to the open Saharan grazing lands. Research on the Ait Unzar Berber pastoralists of the Dra Valley is in its initial stages. However, it is already possible to note some general trends concerning the group's strategies in coping with a changing environment and adapting to the critical resource' situation.

The Extreme Flexibility of Domestic Groups

Among the Ait Unzar the preferred option is for a multi-resource economy, both diachronically and synchronically. According to changing ecological, economic and political contexts, they alternate periods of greater spatial stability—with priority given to the options of agricultural and wage labour—with periods of higher mobility when priority is accorded to pastoral production. But a similar alternation is often practised in a synchronic way at the level of individual domestic groups: a part of the family is then settled in the oasis—mainly in the village of Tiraf, but also in the urban centre of Tagounite—and another part keeps moving with the herds and tents.

Agriculture and herding are not the only options available in such multi-pronged strategies: wage labour exists and the most commonly practised option is military service—the southern part of the Dra Valley is highly militarised as it is strategically important, both because of the Sahara conflict and the proximity of the Algerian border. Another minor option in terms of wage labour is the occasional employment available in the tourism sector. Compared to the Draoua people, emigration, either abroad or to Moroccan industrial cities, is not really a relevant option for nomads and former nomads. Thanks to the elasticity of the roles of family members, such a multiplicity of resources and their contextual use and combination during critical times seem to enable the Ait Unzar to cope with crises better than their farming neighbours.

The Persistence of an Overarching 'Tribal Identity'

Even though the Ait Unzar taqbilt is a cultural and economic unit, it somehow shares this feature with other Nit Atta pastoralists. Nonetheless, even if the Ait Unzar do not consider their unity as founded on genealogical relations—the only proved consanguinal relations are within the five to six generation-deep lineage (ikhs in Berber, fakhda in Arabic)—the group is still conceived of by others and itself as consisting of a sort of more significant territorial and political unit. As already noted, apart from flexibility in accepting 'foreigners' in times of resource scarcity, territoriality does function, with grazing lands and perennial wells being considered the exclusive property of Ait Unzar people.

Moreover, the taqbilt members still recognise themselves as holding exclusive political authority for managing resources and setting conflicts between themselves and their neighbours. In this sense, the presence of a single sheikh for the whole tribe is also significant: this pre-colonial institution has been integrated into the modern national administrative system, but in most cases the sheikh is accorded only territorial recognition, but no genealogical relation with the tribesmen. For example Sheikh Hamman, the eldest son of the former Ait Unzar chief before and during colonisation is acknowledged by the state as sheikh of the village of Tiraf; he is also sheikh of every Ait Unzar between the Algerian border and the Foum Zguid region.

He lives in Tiraf, which is the fixed pole of reference for a territory where Ait Unzar have priority rights to water resources and pastures lands. The tribal assembly representing the interests of all the taqbilt also seems to be active, both in decision-making concerning only the Ait Unzar and in the 'bipolar' articulation with the qabila of settled Draoua people in Tiraf. During crises, this persistence of tribal identity and of some of the main functions of collective tribal institutions even for highly dispersed groups, functions as a backdrop for ensuring wider solidarity in adaptation strategies.

The Change in the Division of Labour

One of the most visible features in this complex interaction of traditional structures and current processes of change is the increasing absence for long periods of adult men from most families, and the 'multi-functionality' of women who occupy the men's roles. Comparing the present situation with past patterns, we find transformations of inner domestic relationships and of the organisation of the household. For example, today it is not unusual to see women going to markets, and this seems related more to the constraints and crisis of modern economy than to the legendary freedom and autonomy of Berber women.

This capacity of women to ensure, not only the functioning of the traditional domestic domain, but also of pastoral production, is sometimes evoked by Ait Unzar men as one of the reasons for maintaining strict endogamy which averages 83 percent for intra-tribal marriage (within the Ait Unzar community) and approaches 100 percent if we include marriages between Berber and Arab pastoral communities.

An initial hypothesis could be that groups such as the Ait Unzar who have not submitted to the pressures to fully sedentarise—pressures built in to the long lasting policies of pre-colonial, colonial and post-colonial regimes—maintain both a wider territory of reference and a sedentary pole of attachment and have a better chance of coping with the critical situation of drought than do settled groups of farmers.

The verification of this hypothesis could add concrete evidence to the debate about the meaningfulness and legitimacy of the widespread policy of forced sedentarisation of nomads implemented by nation states and dictated by international agencies during the previous decades. It could also lead to rethinking future strategies for approaching drought situations in arid zones.

The complementarity between nomadic pastoralists and sedentary farmers in the southern Dra Valley shows once more, that in spite of the ideological weight of an 'anti-nomad' prejudice, there is no clear-cut dichotomy between the nomad and sedentary poles. The definition of Ait Unzar as ruhhala, 'nomads', is still widely used, not only as self-definition but also by settled people and even by state institutions to define communities such as the Ait Unzar. Even such communities and families who have owned permanent houses and cultivated lands in Tiraf since the time of their grand-fathers, can continue to be classified in the ruhhala category, indicating that this term implies more than simple spatial mobility—it carries with it a set of political and social connotations that go to make a complex identity. The Ait Unzar illustrate once more the strong adaptive resourcefulness of nomadic peoples and their resilience in the face of pressures exerted by powerful states and the global economy to make them disappear as 'nomads'.

Antimicrobial Resistance in Escherichia Coli Isolated in Wastewater

Escherichia coli is a major pathogen of widespread importance in commercially produced poultry, contributing significantly to economic losses in both chickens and turkeys and, simultaneously, an important commensal bacteria that inhabits the gastrointestinal tracts of humans and animals. Surveillance of resistance in avian E. coli has a triple benefit: 1) it generates data that support treatment strategies for commercial poultry and thus reduces both the incidence and mortality associated with avian colibacillosis; 2) it offers an "early warning system" with respect to the drug resistance of zoonotic bacteria; and 3) it helps to assess possible human health consequences, since multiresistant nonpathogenic E. coli may transfer their resistance to intestinal bacteria in humans, particularly in individuals undergoing antibiotic treatment.

This may increase the frequency and duration of infections and hospitalisations. Molecular analysis of bacterial genes of drug resistance and of genetic vectors responsible for assembly and mobility of these same genes has demonstrated that identical elements occur in E. coli found in animals and E. coli found in humans.

Drug-resistant avian E. coli may be transferred to humans either through direct contact or, more frequently, through consumption of poultry meat. There is less evidence with respect to the spread in aquatic and terrestrial environments of resistant E. coli, namely evidence of bacteria present in wastewater and sludge from poultry slaughterhouses. Wastewater treatment plants are designed mainly to retain solids, and the quantity and quality of microbiota released into waterways and soils are rarely monitored. From these poultry wastes, E. coli may find ways of spreading to humans either directly, through drinking water or recreational contact, or indirectly, by consumption of meat, fish, and vegetables.

The World Organisation for Animal Health has asked for programs aimed at the monitoring of antimicrobial resistance in food-producing animals. Poultry production is both an important origin of meat for human nutrition and highly dependent on antimicrobial drugs for therapeutic and non-therapeutic (growth promotion and disease prevention) uses. The aim of our study was to assess the number and the resistance patterns of E. coli strains present in wastewater and sludge collected in wastewater treatment plants of poultry slaughterhouses in Portugal.

Wastewater Treatment Plants

Our study collected samples from the wastewater treatment plants of eight Portuguese poultry slaughterhouses (A to H) that were randomly selected from the 40 poultry slaughterhouses that exist in Portugal.

These eight units are responsible for slaughtering 20 percent of the Portuguese broiler production and are located in the two main poultry production areas of the country. Technical data from these plants were recorded: An average of 206,000 broilers were slaughtered per day, corresponding to an effluent emission of 2,341 [m^3] for the receiving rivers and an undetermined volume of sludge that could either be used as soil fertilizer or be disposed of as waste by burial in a landfill after byproduct treatment.

Biological treatment of wastewater was quite heterogeneous: 1) lagoons (plants A, E, G, and H); 2) activated sludge in aerobic conditions (plants B and F); and 3) activated sludge in aerobic conditions followed by a tertiary nitrification treatment (plants C and D). The poultry population consisted of two groups, categorised according to the rearing system used: 1) Conventional broiler flocks were reared in high stocking density (14 to 20 broiler/[m2]) until 30 to 40 days of age, and 2) free-range broilers were reared in open broiler houses at low stocking density (<12 broiler/[m2]), with a minimum age at slaughtering of 81 days. The maximum flock size was 4,800 broilers, and the use of antimicrobial growth promoters has not been permitted since 2000. The plants exclusively slaughtered poultry produced by one or the other of these rearing systems: Plants A to F slaughtered conventionally raised broilers, and plants G and H slaughtered free-range broilers.

Sample Collection

From each plant, wastewater samples of two types were collected in sterile bottles: 1,000 mL from incoming raw wastewater (inflow) and 1,000 mL from treated wastewater (effluent). Each sample was collected during a 4-hour period, with 250 mL harvested every 60 minutes. Four subsamples of sludge (250 g) were also collected (after sewage-sludge press, from storage silos, or from the drying tank) in six of the eight wastewater treatment plants. The subsamples from each plant were later pooled together. All samples were refrigerated and transported to the laboratory for immediate processing.

Sample Processing: Isolation and Enumeration of E. Coli

Tergitol BCIG agar (Biokar Diagnostics, Beauvais, France) (Tergitol) was used for E. coli isolation. Enrichment was performed in both buffered peptone water (Oxoid, Basingstoke, United Kingdom) (BPW) and brilliant green bile 2 percent broth (Oxoid) (BGB), incubated for 12 hours at 37[degrees]C, and streaked for isolation onto Tergitol. After 24 hours at 37[degrees]C, a maximum of 26 E. coli colonies from each sample were selected on the basis of colony size and morphology. Species identification was made with conventional biochemical tests: ability to ferment lactose and glucose (Kligler iron test), to produce hydrogen sulfide, and, when in doubt, to convert tryptophan into indole (indole test).

Volumes of 0.1, 1, 10, and 100 mL from each affluent and effluent wastewater sample were diluted in duplicate in 100 mL of BPW. Each dilution was filtered through a membrane filter with 0.45-[micro]m pore size (Pall Corporation, Michigan), which was placed on Chromocult tryptone bile X-glucuronide agar (Biokar Diagnostics) (TBX). Enumeration of E. coli in sludge samples was performed by agar incorporation in TBX. After incubation at 44[degrees]C for 24 hours, all blue coloured colonies were counted as "presumptive" E. coli (Manafi, 2000). Counts were expressed as colony-forming units (CFUs) per millilitre or per gram of sample.

Antimicrobial Susceptibility Testing

Susceptibility to antimicrobial agents was tested by means of the disk diffusion method according to the National Committee for Clinical Laboratory Standards (NCCLS) guidelines (NCCLS, 2000) for the following antimicrobials:

amoxicillin/clavulanic acid (AMC, 30 [micro]g), ampicillin (AMP, 10 [micro]g), apramycin (APR, 15 [micro]g), cephalothin (CEF, 30 [micro]g), chloramphenicol (CHL, 30 [micro]g), enrofloxacin (ENR, 5 [micro]g), gentamicin (GEN, 10 [micro]g), kanamycin (KAN, 20 [micro]g), nitrofurantoin (NIT, 300 [micro]g), streptomycin (STR, 10 [micro]g), tetracycline (TET, 30 [micro]g), and trimethoprim/sulfamethoxazole (SXT, 25 [micro]g), all supplied by Oxoid. Reference strain E. coli ATCC 25922 was used as a control strain. Interpretation of the diametre of the inhibition zone was made according to the recommendations of NCCLS (2002) and the manufacturer with respect to apramycin (Elanco Animal Health, Greenfield, Indiana). Organisms considered intermediate by this method were recorded as sensitive.

Evaluation and Statistical Analysis

Methods for clustering analysis were applied to establish the associations among the resistance profiles obtained for each isolate. The results for each antimicrobial drug were coded in binary data (0 for susceptible and intermediate phenotypes and 1 for resistant ones), and we computed the corresponding matrix using NTSYS-pc software, applying the Jaccard similarity coefficient and the unweighted pair group method with arithmetic mean (UPGMA) for clustering and dendrogram construction. We used Chi-squared tests to investigate independence between several features of treatment plants and antimicrobial resistances.

Results

The average E. coli loads (expressed in CFUs) per day were 1.35 x [107] per 100 millilitres of raw inflow wastewater and 5.15 x [105] per 100 mL of treated slaughterhouse effluent, respectively. Reduction in the number of E. coli due to the treatment in wastewater plants was variable, ranging from less than 0.5 log to 3 log. For Plant C, the CFU counts in the effluent water were higher than those in the inflow. Plants that used activated sludge treatment in aerobic conditions (B and F) were among the most effective, with 33 and 310 E. coli cells detected in 1 mL of outflow wastewater collected, respectively.

A total of 549 E. coli isolates were obtained: 26 isolates from each inflow and effluent samples and a variable number from sludge samples (the variation was due to the high density of the accompanying flora). E. coli confirmation rate for each sample ranged from 80.8 percent to 100 percent (mean was 92.6 percent).

Chi-squared tests showed no significant differences ($p > .05$) in resistance to any microbial agent by origin of sample (inflow, sludge, or effluent). Furthermore, no significant differences were found between resistance rates and isolation procedures (enrichment with BPW, enrichment with BGB, or no enrichment). Comparison of resistance rate between conventional and free-range broilers revealed significant differences ($p < .05$) only for tetracycline. Nevertheless, E. coli isolated from poultry from conventional rearing farms showed higher drug resistance rates for all antimicrobials, with the exception of amoxicillin/clavulanic acid and nitrofurantoin.

Overall, 106 different E. coli drug resistance profiles were found. In inflow, effluent, and sludge samples, the mean diversity rate of E. coli with respect to resistance profiles was 50.7 percent, 52.9 percent, and 66.2 percent, respectively. Sixty-six isolates (12.0 percent) were susceptible to all the antimicrobials tested, 88 (16.0 percent) were resistant to one active compound, 89 (16.2 percent) showed a double-resistance phenotype, and 306 (55.7 percent) were multidrug resistant (i.e., they had resistance to three or more antimicrobial classes). Resistance to ampicillin and tetracycline alone was commonly observed, whereas resistance to amoxicillin/clavulanic acid, kanamycin, gentamicin, apramycin, chloramphenicol, and nitrofurantoin was constantly associated with multidrug-resistant phenotypes. Forty-seven strains (8.6 percent) exhibited simultaneous resistance to at least 6 of the 12 antimicrobials that were tested; they corresponded to 29 different resistance phenotypes, always encompassing resistance to ampicillin and tetracycline. With one exception, all of these strains were sensitive to amoxicillin/clavulanic acid. Two E. coli isolates showed resistance to 10 antimicrobial drugs. Among 54 isolates with cephalothin resistance, 49 (90.7 percent) were also resistant to ampicillin, but only 15.9 percent of ampicillin-resistant E. coli were resistant to cephalothin. The [AMPR] [CHLR] [TETR] [STRR] [SXTR] phenotype was found in 41 isolates.

We elaborated a dendrogram with resistance profiles in order to look for correlation among the antimicrobials tested. The dendrogram revealed that the most closely related compounds were tetracycline, ampicillin, trimethoprim/sulfamethoxazole, and streptomycin (near 50 percent similarity) (Cluster 1). Gentamicin and apramycin were also clustered together, with a more than 50 percent degree of similarity (Cluster 2). Microbial analysis of E. coli present in waste products of poultry slaughterhouses revealed widespread antimicrobial drug resistance in the more than 500 bacterial isolates that were identified.

Although sub-inhibitory concentrations of antimicrobials and their metabolites, often detected in sewage, might promote growth of resistant bacteria or gene transfer between bacteria in the sewage treatment tanks or in the biofilm lining of the pipelines, no significant differences were found between inflow and outflow resistance rates. Eccles, Searle, Holt, & Dennis (2004), also found small differences in the proportion of drug-resistant populations between final effluents and raw wastewater samples. Interestingly, we found that the poultry-rearing system influences drug resistance rates of E. coli. Conventional approaches to poultry production led to a greater risk that multidrug-resistant E. coli would develop. This result may be due to the more extensive use of antimicrobial agents in conventional rearing houses, which have a high density of stocked birds. Ojeniyi (1985) found all 3,444 E. coli isolates from battery hens to be multiresistant but none of the 2,284 isolates from free-range chickens.

The frequent isolation of multiresistant strains and the overall resistance rates found in our study were noteworthy. Unfortunately, precise figures on the extent and amount of antimicrobial agents used in animal production are not publicly available in Portugal, because no comprehensive reporting system exists for the quantity of antimicrobial agents used in food animals. This lack of accurate information may complicate a definitive interpretation of our data, but on the other hand, it may enhance the usefulness of the data since antimicrobial resistance rates suggest a rough estimate of antibiotic quantities administered to poultry. The dendrogram provides additional information on associations of resistance genes of E. coli and allows the prediction of cooperative selection phenomena with respect to antimicrobial agents. In fact, the rate of E. coli resistance to streptomycin, an antibiotic that is no longer in use, is still high because of selection through frequent use of tetracycline, ampicillin, and trimethoprim/sulfamethoxazole.

This clustering could also be the driving force that maintains E. coli resistance to chloramphenicol despite the ban of this drug for use in animal production since the early nineties. Further genetic studies, namely on Class 1 integrons present in these E. coli isolates, should be performed in the near future. The finding of a high frequency of E. coli strains with tetracycline resistance is consistent with extensive and long-term use of tetracyclines. We found 100 percent resistance in the largest slaughterhouse (Plant A) included in this study, which had declared a systematic use of chlortetracycline in poultry feed between the 18th day and the 21st day to prevent digestive problems and enhance feed efficiency.

Resistance to nitrofurantoin was lower than expected, given the documented illicit use of nitrofurans in Portugal by poultry producers in 2002-2003. This apparent contradiction could be explained by a fast decline of nitrofurantoin resistance in the E. coli population or by insufficient selective pressure since only vestigial levels of these compounds were detected by Portuguese authorities in the majority of poultry samples.

Enrofloxacin and norfloxacin are extensively used in Portugal because their price is low and they have better antimicrobial action against pathogenic E. coli than do more traditional empiric agents (i.e., ampicillin and tetracycline). Nevertheless, the enrofloxacin resistance rate was still moderate in our E. coli isolates. Antunes, Reu, Sousa, Peixe, and Pestana (2003), however, reported 50 percent enrofloxacin resistance in Salmonella sp. isolated from poultry products (60 percent positive) available to consumers in Portugal. It must be added that almost 95 percent of the enrofloxacin-resistant E. coli isolates in our study had an inhibition zone diametre of less than 10 mm. This is a key issue; it is important to detect these highly resistant strains because they are related to the occurrence of stepwise mutations in gene coding for target molecules of quinolones.

In the near future, no new classes of antimicrobial agents are expected to become available for veterinary medicine. This trend, coupled with the revealed data, creates an atmosphere of apprehension, calling for immediate action to minimise further development of antimicrobial resistance in E. coli originating in poultry. Transfer of poultry-originated antimicrobial-resistant E. coli to humans is a potential problem in public health. Our investigation also shows that wastewater treatment processes applied to poultry slaughter waste products are unable to retard the dissemination of resistant E. coli into aquatic environments. Surface water resources represent an important fraction of public drinking water, and standard treatment processes may not be completely efficient in removing or inactivating all infective organisms. Field application of sludge as fertilizer represents another important pathway, via farmland soils, by which resistant E. coli could migrate to ground and surface waters. In conclusion, our investigation documents the presence of numerous multiresistant E. coli strains in the waste products of poultry slaughterhouses in Portugal and shows that these bacteria are not prevented from reaching the general environment by treatment in wastewater plants.

Cryptosporidium Inactivation by Low-Pressure UV in a Water Disinfection Device

Using animal infectivity tests, the authors evaluated a water disinfection device, UV Waterworks [TM] (UVW), for its ability to inactivate cryptosporidium parvum oocysts. The UVW employs low-pressure, germicidal ultraviolet (UV) light, delivering a dose of approximately 120 millijoules per square centimetre (mJ/[cm^2]) under ideal water conditions at a flow rate of 4 gallons per minute (gpm). Dechlorinated tap water containing live oocysts was passed through the UVW at 4 gpm. The oocysts were captured on a filter, separated from the filter, and concentrated into inocula—10 microlitres ([micro]L) each, containing between 103 and 107 oocysts—which were administered orally into 60 neonatal mice. After one week, the mice were killed, and sections of their terminal ilea were analysed microscopically for signs of Cryptosporidium infection. In spite of the high dose of oocysts, none of the mice showed signs of infection. A process control run with the UV lamp off resulted in 95 percent infection at a dose of 103 oocysts per inoculum. The calculated reduction in oocyst infectivity from passage through UV Waterworks was at least 5.4 orders of magnitude. The authors conclude that exposure to low-pressure UV at 120 mJ/[cm^2] effectively disables Cryptosporidium.

Through NEHA's long-standing relationship with NSF International, NEHA was granted permission by NSF International to share with the Journal's readership various papers that were presented January 12-15, 2000, at the "NSF International Small Drinking Water and Wastewater Systems International Symposium and Technology Expo" in Phoenix, Arizona. This chapter, "Cryptosporidium Inactivation by Low-Pressure UV in a Water Disinfection Device," is one of them. It is important to note that these papers were screened by an NSF International advisory committee prior to their presentation at the conference, but they have not been peerreviewed by NEHA's Journal program for technical accuracy.

Because these papers contain useful and interesting ideas and information that might be either delayed or lost if the papers were sent through the Journal's normal peer review process, NEHA has decided to publish them as presented, with only minor editorial modifications. We hope you look forward to more of these papers in future issues of the Journal!

Certain pathogenic organisms, such as Cryptosporidium parvum, evade removal, resist chlorine disinfection, or do both in some municipal drinking-water treatment systems. They thus become a threat to immuno-compromised individuals consuming tap water. Ultraviolet (UV) disinfection is among the methods being investigated by researchers for treatment of municipal tap water to deactivate Cryptosporidium. In this chapter, the authors report new results showing that low-pressure UV can be highly effective at inactivating this organism. Cryptosporidium Viability Versus Infectivity A study that uses human hosts is the most accurate method of assessing the effect of any disinfection method on the infectivity to humans of Cryptosporidium parvum oocysts. In practice, since such studies entail severe financial, legal, and ethical burdens, animal surrogates are used. Even animal studies are often, however, prohibitively expensive and time-consuming. Thus, several other methods have been developed to gauge the degree to which various disinfection treatments inactivate Cryptosporidium oocysts.

Two methods have been in common use in the past few years: the DAPI/PI (4',6'-diamidino-2-phenylindole and propidium iodide) vital-dyes assay and excystation. These methods can demonstrate the viability of oocysts, and nonviable oocysts can be considered noninfective. A viable oocyst is not, however, necessarily infective. As explained below, equating viability with infectivity has in the past led to the false conclusion that UV light is relatively ineffective against Cryptosporidium.

Cryptosporidium and UV Disinfection

Initial studies of the effect of UV light on Cryptosporidium oocysts reported in the literature used only the vital-dyes and excystation methods. It was concluded that UV doses on the order of 10,000 mJ/[cm^2] were needed to achieve a 2- to 3-log inactivation. Pulsed, medium-pressure UV was later shown to result in a 2- to 3-log inactivation at around 2,000 mJ/[cm^2], but a conventional, low-pressure UV system delivering 180 mJ/[cm^2] was shown not to reduce the viability of oocysts significantly.

In spite of the disappointing results on the reduction in viability, several researchers recently have tested the effect of low doses of medium-pressure UV on the infectivity of oocysts. The results were surprising, showing a 3.9-log inactivation at only 19 mJ/[cm^2] and a greater-than-4-log inactivation at 41 mJ/[cm^2]. Unpublished results of collimated-beam experiments using low-pressure UV and animal infectivity in early 1999 indicated that greater-than-4-log inactivation could be achieved at only 16 mJ/[cm^2]. It was due to these encouraging results that the mouse infectivity assays reported here were performed.

Description of UV Waterworks [TM] (UVW)

UV Waterworks (UVW), a product of WaterHealth International, Inc., is an unpressurized disinfection unit designed for flow rates of up to 4 gallons per minute (gpm) (15 litres per minute [L/min]). Disinfection is achieved with a low-pressure mercury germicidal lamp suspended above an open-channel flow of water. Water flows through the unit by gravity. The mean residence time of water in the unit is 12 seconds at 4 gpm. The unit uses 60 watts of power and produces a UV fluence (or "dose") in the water of approximately 120 mJ/[cm^2] under ideal conditions (water with turbidity less than 1 nephelometric turbidity unit [(NTU)] and UV transmittance greater than 95 percent at 1 cm). The fluence under such conditions was determined through biodosimetry with Bacillus subtilis spores according to the procedure outlined in Annex B and Annex C of NSF Standard 55 (NSF, 1991).

The purpose of this experiment was to measure the effect of UVW on Cryptosporidium parvum oocyst infectivity with neonatal mice as hosts. To isolate the effect of UV light from possible influences of the physical passage of oocysts through the UVW, a process control was run in which the UV lamp was turned off. A trip control, consisting of oocysts that were not exposed to any of the experimental procedures prior to preparation of the mouse inocula, was also run to ensure that oocyst infectivity was not being reduced significantly by any extraneous factor.

Test Setup

Tap water was filtered through 1.0-and 0.5-micron ([micro]m) thermally bonded polyolefin fibre filters (Corning CoaX [TM]), and any residual chlorine was allowed to degas to the atmosphere over a period of three days. The turbidity was measured to be less than 0.1 NTU. The transmittance of 254 nanometres (nm) UV was 95 percent over a path 1 cm long. The filtered, dechlorinated water went into the 60-litre feed tank. The bottom delivery port of the tank was connected to the UVW inlet via a centrifugal pump and a valved flow metre. Just upstream of the UVW inlet, a spike injection port equipped with a check valve was incorporated into the line feeding into the UVW. This port was connected to a graduated spike cylinder via a metring diaphragm pump. The effluent from the UVW was directed into a 90-L capture tank. A self-priming centrifugal pump sucked water from the capture tank through a 1.0-[micro]m (absolute) polycarbonate track-etch membrane filter (Corning Nuclepore [TM]) held in a capture filter housing 293 millimetres (mm) in diametre and discharged the filtered water to the drain.

Oocyst Exposure to UV and Process Control

Cryptosporidium parvum oocysts of the Harley Moon Isolate (National Animal Disease Centre, Ames, Iowa) were obtained fresh (less than one month old) from the University of Arizona. This particular isolate had been maintained by passage through neonatal calves. A pre-enumerated stock solution containing 1 x [108] oocysts per mL was divided into two equal 10-mL portions. One portion was assigned to the UV exposure test.

A 100-[micro]L aliquot was withdrawn from the second portion to serve as the trip control. The rest of the second portion was reserved for the process control. The vial containing the pre-enumerated oocysts (one for the UV exposure, another for the process control) was vortexed for 30 seconds and emptied into the spike cylinder. The vial was then rinsed with filtered, dechlorinated tap water and emptied into the cylinder three times. Filtered, dechlorinated water was added to the cylinder to a total volume of 100 mL and thoroughly mixed.

Ten minutes before the start of the UV exposure test, the UV disinfection unit was turned on so that the UV lamp would warm up to its normal operating temperature. For the process control, the unit was not turned on (electrical power was not needed for the water to pass through, only for the UV lamp to be on). The system feed pump then began pumping water into the UVW from the feed tank at 4 gpm (15 L/min). When the flow was established, the diaphragm pump began pumping the 100-mL spike dose of oocysts into the feed stream at 200 mL/min. After the spike cylinder was emptied, additional volumes of filtered water were added over the course of two minutes to rinse the cylinder and spike plumbing. The main flow of water to the UVW was continued for one minute beyond the termination of the spike system rinsing to ensure that close to all oocysts were flushed out of the UVW.

The treated (UV-exposed or process-control) oocysts and water were collected in the capture tank. The centrifugal pump drew the entire tank volume plus a deionized water rinse of the tank walls through the capture filter. The filter membrane, with the trapped oocysts on its surface, was removed from the filter housing and placed in a beaker containing 100 mL of buffered solution. After the membrane was washed in this solution for two minutes by hand, was sonicated for five minutes, and was again washed by hand for two minutes, it was removed and discarded. The solution remaining in the beaker contained the treated oocysts. It was centrifuged at 1,050g and aspirated to a l-mL pellet. This pellet was chilled and shipped overnight to the infectivity laboratory. It was kept chilled up to the time of the initiation of the infectivity assay, which was within 24 hours of the UV exposure.

Mouse Study

A bright-field hemacytometre was used to determine the concentration of oocysts present in the three vials containing the oocysts from the UV exposure, the process control, and the trip control, respectively. These concentrations were used to calculate and prepare the dilutions necessary to achieve the desired dose of oocysts in each of the mice inocula. The inocula were prepared from serial dilutions of the pre-enumerated starting suspension with sterile water adjusted to pH 7. Calibrated pipettors were used for all dilutions, Six sub-sample counts of the diluted oocysts were made to verify the concentration fed to the mice. The counts were performed and cross-checked by two technicians.

The mice used for this study were four-to-six-day-old neonatal CD-1 outbred mice. Approximately 60 mice were used in each of the three study conditions (trip control, process control, UV exposure). Of these 60, approximately 20 received a low dose of oocysts, 20 received a medium dose, and 20 received a high dose. To make possible detection of a potential reduction in oocyst infectivity of up to six orders of magnitude, target doses of the UV exposed oocysts were [103], [105], and [107] per inoculum. The target doses of the process-control oocysts were [102], [103], and [104] to allow for a 1- to 2-log reduction in infectivity due to factors other than UV light. Trip-control oocysts were expected to show very little loss in infectivity. Thus, their target doses were chosen to span the useful range of the reference dose-response function, which relates the dose of infective oocysts to the fraction of mice that become infected.

The oocysts were administered orally in 10[micro]L doses with a dedicated, calibrated pipettor equipped with a standard tip. The mouse was held in the palm of the technician's hand until the entire dose had been ingested. After seven days the animals were killed, and approximately 2 cm of each terminal ileum was excised. The tissue samples were fixed in formalin, embedded in paraffin, sectioned, mounted on microscope slides, and stained. The stained samples were then examined for evidence of Cryptosporidium parasite life cycle stages in the borders of the intestinal villi. Specimens with parasites were scored positive, and those without parasites were scored negative.

Infectivity Calculations

The proportion of mice infected in each case was used in conjunction with the reference dose-response function to calculate the number of infective oocysts in the inoculum. The model was determined previously from experiments dedicated to this purpose. The trip control showed a reduction in oocyst infectivity of between 0.3 and 0.6 logs. The reduction might be due to some aspect of the handling, transport, or storage of these microorganisms from the time they were collected to the time they were administered to the mice. The other possibility is that the apparent low rate of infectivity reflects a difference between the mice used in this study and the mice used to generate the Korich et al. (1999) model—that is, a manifestation of the natural variability in the susceptibility of mouse populations.

As expected, the process control showed a slightly higher loss of infectivity overall than the trip control. This result is probably due to the more numerous environmental changes that the process-control oocysts experienced during the experiment. Nevertheless, at an inoculum dose of 1,087 oocysts, 95 percent of mice were infected, and at an inoculum dose of 9,500 oocysts, 100 percent of mice were infected. By contrast, the UV-exposed oocysts were not able to produce infection in any mice at any of the three doses. If a single mouse had been infected at the highest oocyst dose, the reduction in oocyst infectivity would have been six orders of magnitude.

Cryptosporidium parvum oocysts that were passed through a UV water disinfection device delivering approximately 120 mJ/[cm^2] were effectively disabled. The UV-treated oocysts did not produce infection in any of the 60 mice to which they were administered. From the number of oocysts in the inoculum, the authors observed a reduction of greater than six orders of magnitude in the number of infective oocysts. When the maximum log reduction resulting from the three trip-control trials was subtracted, the effective log reduction from passage through the UVW alone would be at least 5.4 logs.

These results demonstrate that low-pressure UV at fairly low doses can be as effective against Cryptosporidium oocysts as it is against most bacteria and viruses. Previous results suggesting the contrary were limited because they came from studies of viability that used only staining or excystation techniques. There is now ample evidence that lack of oocyst viability as measured by these techniques is not a good indicator of infectivity. More studies like this one are needed to assess the effectiveness of UV disinfection against other harmful cysts and protozoans.

10

Outbreaks in Drinking-Water Systems

For an incident to be classified as a waterborne outbreak, at least two people must have experienced a similar illness after the ingestion of drinking water or after exposure to water used for recreational purposes, and epidemiological evidence must implicate water as the probable source of the illness. An exception is made for a single case of chemical poisoning (e.g., methemoglobinemia) when water-quality data indicate contamination by the chemical. The surveillance system reports outbreaks associated with drinking water, recreational waters (swimming pools, water parks, and fresh and marine surface waters), and whirlpool- and hot tub—associated outbreaks of dermatitis, but only outbreaks related to drinking water are included in the analysis here. Although most of the reported outbreaks were associated with water used or intended for drinking or domestic purposes, several were associated with the ingestion of water not intended for consumption (e.g., the use of springs and creeks by backpackers and campers) and miscellaneous deficiencies, such as water contaminated at its point of use (e.g., water faucets, serving containers, and ice).

Public-water systems are regulated by U.S. EPA and are classified as either community or noncommunity systems. A community water system serves year-round residents of a community, subdivision, or mobile-home park that has 15 or more service connections or serves an average of 25 or more residents. A noncommunity water system is used by the general public for 60 days or more per year and has at least 15 service connections or serves an average of 25 or more people. Of the approximately 170,000 public-water systems in the United States, 55,000 (32 percent) are community systems and

115,000 (68 percent) are noncommunity systems. Community water systems serve some 243 million people. Millions of people use noncommunity systems while travelling or working in the United States. Some 24 million people (9 percent) rely on individual water systems, which are not owned or operated by a water utility and serve less than 15 connections or less than 25 people. The responsibility for regulating individual water systems rests with the state or county.

Outbreak Statistics

During 1991-1998, 126 drinking-water outbreaks and more than 429,000 cases of illness were reported in 41 states and three U.S. territories. Most outbreaks (87 percent) were reported in public water systems. Sixty-two outbreaks (49 percent) and 6,562 illnesses were reported in noncommunity water systems. Forty-seven outbreaks (37 percent) and 422,298 illnesses were reported in community water systems. A single outbreak of cryptosporidiosis in Milwaukee was responsible for an estimated 403,000 cases of illness, 440 hospitalisations, and 50 deaths in 1993.

This outbreak is the largest reported in the United States since the collection of statistics began in 1920. Seventeen outbreaks (14 percent) and 161 illnesses were reported in individual water systems and among persons who consumed water not intended for drinking.

More than 156,000 public systems (92 percent) rely primarily on groundwater sources, and slightly more than 13,000 systems (8 percent) use surface water. Except for 11 public water systems where the outbreak was caused by contaminated ice, water taps, or storage containers, or where the water source was not identified, water systems were classified either as surface water or as groundwater. Seventy-four (68 percent) of the outbreaks occurred in public systems that used groundwater; 24 outbreaks (22 percent) occurred in surface-water systems.

A bacterial, viral, or protozoan etiology was identified in 51 outbreaks (41 percent). In 52 outbreaks (41 percent), an infectious etiological agent was suspected but the etiology was not identified. Of the 31 community-system outbreaks with a known or suspected infectious etiology, 15 (48 percent) occurred in surface-water systems and 13 (42 percent) in groundwater systems. Of the 61 infectious-disease outbreaks reported in noncommunity systems, 52 (85 percent) occurred in groundwater systems, and only two (3 percent) occurred in surface-water systems. Eight (73 percent) of the 11 infectious-disease outbreaks in individual water systems occurred in groundwater systems.

Chemical contaminants were responsible for 23 (18 percent) of all outbreaks; 16 outbreaks (34 percent) occurred in community systems. Most community-system chemical outbreaks (69 percent) were associated with distribution system contamination. High copper levels due to corrosive water were identified in seven outbreaks (44 percent). Cross-connections and back-siphonage (nitrate, soap concentrate, and fluoride) caused four outbreaks (25 percent). Five outbreaks (31 percent) were associated with inadequate control of chemicals added during water treatment (fluoride, sodium hydroxide, and chlorine).

Only one chemical outbreak (copper) was reported in a noncommunity water system, but six of the outbreaks (35 percent) in individual systems were associated with chemicals. Three single-case outbreaks of methemoglobinemia were due to high levels of nitrate in well water. In one outbreak, nitrate levels exceeded maximum limits even though a reverse-osmosis membrane filter had reduced levels in the water source.

Three single-case outbreaks were detected through a leadscreening program. Water at the homes of infants with high blood lead levels was corrosive, leaching lead from fittings and lead-soldered seams in water storage tanks. Because of the large outbreak in Milwaukee, almost all (99 percent) of the illnesses in community-system outbreaks occurred in surface-water systems. Outbreaks in noncommunity systems caused 6,562 illnesses, almost all (94 percent) of which occurred in groundwater systems.

Water System Deficiencies

For most water systems, an investigation identified the deficiency that likely caused the outbreak. Water treatment deficiencies included inadequate, interrupted, or no disinfection of groundwater; inadequate disinfection of unfiltered surface water; and inadequate or interrupted filtration of surface water. Outbreaks also were caused by contaminated, untreated groundwater and contaminants that entered the distribution system.

Outbreaks in community water systems were associated with distribution system deficiencies (36 percent), inadequately treated groundwater (19 percent), inadequately treated surface water (17 percent), miscellaneous or undetermined causes (17 percent), and inadequate control of chemical feed (11 percent). In noncommunity water systems, outbreaks were associated with inadequately treated groundwater (63 percent), miscellaneous or undetermined causes (23 percent), distribution system deficiencies (13 percent), and inadequately treated surface water (1 percent). Outbreaks in individual water systems were associated with inadequately treated groundwater (59 percent), distribution system deficiencies (24 percent), miscellaneous or undetermined causes (11 percent), and inadequately treated surface water (6 percent).

Deficiencies Associated with Source Water Contamination

Inadequate water treatment in community systems caused 17 outbreaks (36 percent), most (53 percent) of which were associated with contaminated wells.

Almost all (91 to 97 percent) of the outbreaks associated with source-water contamination and inadequate treatment in individual and noncommunity systems occurred in well water systems. In noncommunity systems, slightly fewer outbreaks (46 percent) were associated with contaminated, untreated well water than were caused by inadequate and interrupted disinfection of well water (54 percent). Two outbreaks were associated with deficiencies in UV-light disinfection systems, and in the remaining outbreaks, chlorine disinfection was either inadequate or interrupted.

Sources of contamination were identified in 71 percent of the outbreaks associated with untreated well water. Sewage contamination of wells was identified in eight out-breaks (26 percent). Other infectious-disease outbreaks were associated with wells drilled in limestone and fissured rock (16 percent), wells vulnerable to surface-water runoff or floods (13 percent), and improperly constructed or located wells (10 percent).

Distribution System Deficiencies

In community water systems, distribution system contamination caused as many outbreaks as source-water contamination. A distribution system deficiency was identified in 17 (36 percent) of the outbreaks reported in community systems. Outbreaks occurred when contaminants entered the distribution system either through a cross-connection (47 percent) or during storage or main repair (12 percent); corrosive water was responsible for seven outbreaks (41 percent). Although noncommunity and individual water systems generally have small, limited water distribution networks, they are also subject to distribution system contamination and corrosive water. Four (24 percent) of the outbreaks in individual water systems and eight (13 percent) of the outbreaks in noncommunity systems were associated with distribution system deficiencies, primarily cross-connections and corrosive water.

Miscellaneous and Undetermined Deficiencies

Seventeen outbreaks (71 percent) were associated with the use of contaminated ice or water at its point of use, consumption of water not intended for drinking, and contaminated bottled water. These deficiencies are not related to the adequacy of water treatment or operation and should be corrected by food safety regulations and health education. An outbreak of acute gastroenteritis among hikers was associated with the use of water from an outside tap at a wastewater plant. The tap was not marked as nonpotable, and water from the tap had more than 23 fecal coliforms per 100 millilitres (mL). In another outbreak, consumption of water from an outdoor faucet by attendees at a day camp resulted in 77 primary cases of cryptosporidiosis and 24 probable secondary cases in household members. In Saipan, an outbreak of non-O1 Vibrio cholerae was associated with bottled water from two commercial bottling plants. Bottled water tested positive for fecal coliforms, but the source of contamination was not determined. In seven outbreaks (29 percent), a water system deficiency was not identified, primarily because of an incomplete investigation. In an extensively investigated outbreak in Clark County, Nevada, however, no water treatment deficiencies or breakdowns were discovered, and water quality was found to be excellent before, during, and after the outbreak. Although investigators concluded that cryptosporidiosis among HIV-positive adults was associated with consumption of municipal water, there are concerns that the observed waterborne association may be due to recall bias.

The chemical or pathogen responsible for the outbreak was detected in the water source or water system during 21 (91 percent) of 23 chemical outbreaks and 21 (78 percent) of 27 infectious-disease outbreaks in which samples were collected and analysed. Cryptosporidium, Giardia, Shigella, and E. coli O157:H7 were isolated from water samples collected during eight, six, three, and two outbreaks, respectively. Enteric viruses and Salmonella were isolated during one outbreak each. Water samples were analysed for coliform bacteria during the investigation of 81 outbreaks of infectious disease.

U.S. EPA's Total Coliform Rule (TCR) requires public water systems to routinely monitor their tap water for total coliforms. Coliform bacteria are members of the Enterobacteriaceae (e.g., species of Enterobacter, Klebsiella, Citrobacter, and Escherichia) and have been used for decades to assess the microbiological quality of drinking water. These bacteria are present in the normal intestinal flora of humans and other warm-blooded animals and are found in large numbers in fecal wastes. Most species of coliforms are also free-living in the environment, and their presence in drinking water does not necessarily represent fecal contamination. During 50 outbreaks, samples were collected and analysed for fecal coliforms, a marker of human fecal contamination.

Water samples were usually collected during the early stages of an outbreak. In several investigations, samples were not collected until two to four weeks after the beginning of the outbreak. In one outbreak, water samples were collected two months after the beginning of the outbreak. In seven outbreak investigations, water samples were collected from the water sources (wells) of noncommunity systems rather than from the distribution systems. In these seven outbreaks, information from source-water samples was used as a substitute for information from tap water samples. Because the well water was not treated, it was presumed that tap water samples, if they bad been collected, would likely be of similar quality.

During many outbreak investigations, more coliform samples were collected than would be required by the TCR for routine monitoring purposes, and the samples were collected during relatively short periods of time, usually one-to two-week periods. As many as 60-samples (median = 4) and as many as 16 samples (median = 3) were collected during the investigation of outbreaks in community and non-community systems, respectively.

Total coliforms were detected during the investigation of 59 outbreaks (73 percent) in which samples were collected. Coliforms were found in all individual systems and most noncommunity systems (83 percent) but in only 46 percent of the community systems. In 12 noncommunity-, five individual-, and three community-system outbreaks, the only water sample that was collected was found to be coliform positive. In six outbreaks, one coliform-positive sample was found, with two to 20 samples collected from each outbreak. Fecal coliforms were found in 41 (82 percent) of the outbreaks in which water samples were analysed for fecal coliforms.

Total coliforms were detected during the investigation of all 16 outbreaks of bacterial etiology (100 percent) and during the investigation of 35 outbreaks (81 percent) caused by viruses and unidentified agents; however, coliforms were detected during only eight outbreaks (36 percent) caused by Cryptosporidium or Giardia. In community systems, coliforms were detected in five outbreaks caused by protozoa (33 percent). In noncommunity systems, coliforms were not found during the four protozoan outbreaks in which samples were collected, but in individual systems, coliforms were detected during the investigation of all three protozoan outbreaks in which samples were collected.

Outbreaks of Undetermined Etiology

Success in identifying an etiological agent in waterborne outbreaks has improved only slightly in the past 25 years. During 1991-1998, an etiological agent was determined in 59 percent of waterborne outbreaks compared with 45 percent and 55 percent of outbreaks reported during 1971-1980 and 1981-1990, respectively. Investigations are not always timely and clinical specimens are not always collected. Even when specimens are collected, the analysis may not be comprehensive. Although methods for detection of enteric viruses have improved greatly in recent years, diarrheal stool specimens generally are examined for bacterial pathogens, not for viruses. In many laboratories, testing for Cryptosporidium is done only if requested and may not include routine stool examinations for ova and parasites. During 1993-1996, illnesses associated with at least six of 14 outbreaks of undefined etiology had incubation periods, durations, and symptom complexes that were consistent with viral syndromes, but limited or no viral testing was done.

Testing in four outbreaks of suspected viral etiology found that stool specimens were negative for bacterial pathogens, and testing in three outbreaks found that stool specimens were negative for both bacterial and parasitic pathogens. In one outbreak of undetermined etiology in 1995, the state laboratory identified rotavirus in two stool specimens, but no testing was done for Norwalk-like viruses, which have been more commonly associated with waterborne outbreaks. Additional knowledge about etiological agents can provide more information about the effectiveness of regulatory and other efforts to reduce waterborne outbreak risks. Investigators are encouraged to collect clinical specimens and submit them for complete testing either to CDC or to state laboratories that conduct viral and protozoan, as well as bacterial analyses.

Water System

Almost half (44 percent) of all outbreaks in public water systems were caused by the inadequate disinfection of groundwater and use of untreated, contaminated groundwater. These statistics suggest that additional regulations and increased surveillance activities may be needed for groundwater systems. Wells and springs should be better protected from sources of contamination, especially surface-water runoff and underground seepage from septic-tank and other sewage discharges. Periodic sanitary surveys and hydrogeologic assessments can help identify systems with the highest possibility of fecal contamination. Some groundwaters may require disinfection. Disinfection should be adequate in terms of contact time and concentration, not interrupted, and sufficient to maintain a chlorine residual throughout the system.

Distribution system deficiencies were responsible for a significant number of outbreaks in public water systems, especially community systems. Increased attention should be paid to maintaining the integrity of the distribution system and preventing contamination from entering the system through cross-connections and back-siphonage, and during water main repairs and construction. Water system managers also should assess the corrosive potential of their source water, identify possible sources of contamination for distribution system storage facilities, and take appropriate steps to remedy any deficiencies. Chlorine-residual monitoring can help warn of potential contamination. When decreased residual levels are detected, operators should initiate investigations to identify sources of contamination.

Giardia and Cryptosporidium continue to pose waterborne risks, especially in groundwater systems, where 55 percent of 11 outbreaks caused by Cryptosporidium and 40 percent of 15 outbreaks caused by Giardia were associated with groundwater contamination. The inadequate treatment of surface water was responsible for 27 percent of cryptosporidiosis outbreaks and 40 percent of giardiasis outbreaks.

The continued occurrence of protozoan outbreaks in surfacewater systems underscores the importance of meeting U.S. EPA's turbidity standards and other provisions for treating surfacewater sources. It is also important to assess the potential for sewage contamination of groundwater sources, especially seepage from on-site systems and surface runoff. Groundwaters contaminated by protozoa are considered to be under the direct influence of surface water and are subject to U.S. EPA's filtration requirements.

Coliforms in Water During Outbreaks

Coliforms were found during the investigation of all outbreaks of bacterial etiology, but in less than half of the outbreaks caused by Giardia or Cryptosporidium. Similar findings were reported in a previous study of waterborne outbreaks reported during 1983-1992. Although coliforms can be adequate indicators for bacterial pathogens, they are inadequate indicators for the presence of protozoa. Protozoa are more resistant to water disinfectants than are bacteria, and disinfection may be sufficient to inactivate coliforms but not protozoa. As in earlier findings, coliforms were detected during most outbreaks of undetermined etiology. If unidentified viruses caused half of these outbreaks, as suggested by recent outbreaks, coliforms may sometimes serve as an indicator for waterborne viruses. Coliforms may not always, however, indicate the presence of waterborne viruses. Some viruses survive longer in water than do coliform bacteria and are more resistant than coliforms to water disinfectants normally used in water treatment. Also, coliforms were detected during only two of the four viral outbreaks reported during 1991-1998.

Although the total-coliform test may indicate the presence of bacterial pathogens during an outbreak, it is important to remember that coliform analyses during an outbreak investigation may not be relevant in assessing the vulnerability of water systems to an outbreak. Water quality monitoring is often intense during an outbreak investigation, and water samples are being collected during or just after a contamination event. In some instances, contamination was sufficiently large that a single sample was found to be positive for coliforms. In other instances, no coliforms were detected even though a large number of samples (16 to 44) were collected. This finding emphasizes that water contamination sufficient to cause an outbreak can be intermittent and short-lived. Thus, the timing of the sample collection, the number of samples collected, and the selection of appropriate indicators are all important considerations in the development of a monitoring program that can assess a system's vulnerability to an outbreak. For example, under the current TCR minimum-sample-collection requirements, a large or continuous source of contamination would be necessary to detect coliforms in the few samples collected by small systems. Community systems that serve 1,000 or fewer persons are required to collect only a single coliform sample each month. For systems serving 1,001 to 4,900 persons, only two to five samples per month are required. Only for communities of more than 25,000 people are 30 or more samples required each month. Most noncommunity systems need collect only one sample per month or quarter.

Routine coliform surveillance records for 45 outbreaks during 1991-1998 showed that only 22 percent of community and 9 percent of noncommunity systems experiencing an outbreak had violated U.S. EPA's maximum contaminant limit (MCL) for total coliforms in the 12-month period before the outbreak.

This finding suggests that current routine coliform-monitoring programs alone are not able to identify water systems that are vulnerable to an outbreak. Similar findings were reported in an evaluation of the previous TCR. In that study only 38 percent and 19 percent of community systems reporting an outbreak had exceeded the total-coliform MCL during the three and 12 months before the outbreak, respectively. The greater frequency of coliform detection in water systems during an outbreak investigation may be due to more intensive monitoring, insufficient coliform-monitoring requirements, or both.

Most waterborne outbreaks were caused by inadequate or no treatment of groundwater and the contamination of water during its storage and distribution to the customer, especially through cross-connections and corrosion.

Additional regulations and increased surveillance activities, along with appropriate corrective measures, are needed to reduce waterborne-disease risks posed by these sources of contamination. Outbreak data underscore the importance of U.S. EPA's filtration and turbidity requirements for surfacewater systems and for groundwater systems under the direct influence of surface water.

U.S. EPA (1991b) has promulgated regulations to control corrosion by-product, and water utilities should pay more attention to cross-connection control regulations and inspections, monitoring and treatment for corrosive water, and maintaining distribution system integrity. To reduce waterborne-disease risks in groundwater systems, U.S. EPA (1999) has recently proposed requirements for sanitary surveys, hydrogeologic assessments, and disinfection. Protozoan outbreaks in groundwater systems emphasize the importance of assessing the potential for sewage contamination of wells and springs and identifying sources that are under the direct influence of surface water.

Outbreak data suggest that coliform-monitoring requirements are inadequate to identify water systems that are vulnerable to an outbreak. For most public systems that reported an outbreak, the TCR required the collection of only one to three coliform samples each month. The TCR focuses on detection of coliforms in the distribution system after treatment, It may be more effective for groundwater systems to monitor source-water quality and assess the potential for contamination from sewage discharges and surface water.

Although coliforms adequately indicate the presence of bacterial pathogens in drinking water during an outbreak, they are not a good indicator for Cryptosporidiwn, Giardia, and waterborne viruses. Consequently other indicators should be considered. Activities such as frequent sanitary surveys should be conducted in addition to monitoring. Although the reporting of outbreaks is incomplete and the accuracy of case counts varies, waterborne-outbreak-surveillance data have helped identify the types of water systems, their deficiencies, and the respective etiologic agents associated with the outbreaks. Outbreak detection activities and investigations should be improved so that more complete information is obtained about the causes and etiologies of waterborne outbreaks.

Epidemiologists should collaborate with engineers and environmental health specialists during investigations to evaluate water systems and sources of contamination. These data are important for evaluating source water protection strategies, water treatment technologies, and drinking-water regulations and for influencing research priorities.

Diarrheal Diseases in Children from a Water Reclamation

This study was conducted to assess the risk of enteric diseases among children living in a water reclamation area in Mexico City. A geographic information system was used to define eligible wells and surrounding homesteads.

Sixty-five water samples from five wells were tested for fecal coliform bacteria per 100 mL (FC/100 mL) during visits to 750 eligible households; caretakers only in those dwellings with children under 5 years old were interviewed throughout repeated cross-sectional surveys, conducted during 1999-2000.

Data on diarrheal diseases were obtained from 761 children during the rainy season and 732 children during the dry season; their guardians also provided information on drinking water supply, sanitation, and socioeconomic variables. The presence of indicator organisms in groundwater samples pointed to fecal pollution; bacterial indicators, however, did not predict the health risk.

The rates of diarrhea were 10.7% in the dry season and 11.8% in the rainy season. Children 1 year old showed the highest rate of diarrhea during the dry season [odds ratio (OR) = 2.1 with 95% confidence interval (CI), 0.99-4.71], particularly those from households perceiving unpleasant taste of tap water and consuming vegetables washed only with water.

Lower risk was observed in individuals enjoying full-day water supply and a flushing toilet, as well as those storing water in covered receptacles. Rainy season data suggested that children from households perceiving a colour to their water had a higher rate of diarrhea than did those without such complaint; recent consumption of food sold by street vendors was also a significant risk factor. Groundwater is at risk of contamination, as indicated by the presence of FC/100 mL. The endemic pattern of diarrhea, however, reflects mostly inadequate housing, sanitation, and water-related practices. Health protection policy must be discussed.

Water and sanitation deficiencies represent a growing environmental health challenge in several regions around the globe. Unsafe sewage disposal and fecal-oral transmission of pathogens are responsible for otherwise preventable enteric diseases and 3.2 million premature deaths every year. In less developed countries, the disease burden falls heavily on the poor. This gap is perpetuated by the fact that environmental interventions have neglected sanitary needs and focused on the development of drinking water supplies instead. Yet the more dramatic "life-saving" oral rehydration salts therapy (an ethical imperative in primary health care) has shifted the attention from the actual role of prevention to cost-effective "solutions". Exposure to fecal pollution is growing as a result of economic driving forces, overcrowded slums, and weak institutions.

The current population of the Mexico City metropolitan area (MCMA) is 18 million and is forecast to be 23.5 million by the year 2015. Up to 75% of its water supply depends on groundwater reserves. Overextraction of water has led to soil subsidence and cracking of underground pipes, which may facilitate the mixture of drinking water supplies and sewage, as well as the downward migration of pollutants. Earlier investigations indicated high rates of groundwater positive microbiologic tests; despite growing concern, public health data are limited and official reports provide scarcely credible information. Additional gaps reflect the limitations of microbial indicators currently used to assess drinking water quality; therefore, the basis for "safety" criteria stipulated by national regulations is increasingly debatable.

A water reuse program, consisting of wastewater treatment and effluent reclamation (e.g., irrigation of fodder and green belts) is being developed in MCMA. A series of investigations is being conducted to assess the risk of enteric disease and provide some basis for future environmental interventions. This study addressed the following research questions: Is groundwater microbiologic pollution a health risk? What are the risk factors for enteric diseases? And which further questions should be addressed? The boundaries of Mexico City were first framed within basic geostatistical areas, which in turn were characterised according to demographic variables, as provided by the national census tracks. Earlier investigations described the environmental indicators linked to enteric diseases, from which a "high-risk" communities approach was further developed. Our present case study was a second-step approach, resulting from an earlier investigation.

The research area is located in Xochimilco, on the outskirts of the city, where a water reclamation project is being developed. This project consists of a series of wastewater treatment plants, the effluents of which flow through a network of canals to be reused for agricultural irrigation, all of which contributes to the recharge of groundwater reserves, for subsequent extraction (i.e., pumping wells).

Eligible study units were homesteads within 500 m of selected wells. A trained technician gathered a total of 65 water samples from five wells (35 samples during the rainy season and 30 in the dry season). Water samples were collected at a point before chlorination and distribution processing, kept on ice at 4[degrees]C, and transported to the laboratory .

Water samples were incubated 24 hr, and the development of specific colour changes (fluorescence) indicated the presence of fecal coliform bacteria (FC/100 mL), which was tested by using the Colilert method (presence/absence), as described by Edberg et al. and approved by the U.S. Environmental Protection Agency. Contaminated wells were defined as those showing positive results in 95% of their water samples, whereas controls were defined as those wells consistently showing negative tests.

Only households having children under 5 years old were numbered and spatially located via GIS. A random sampling technique was used, and upon previous informed consent, 750 eligible households were included in two repeated cross-sectional studies. A total of 732 children participated in the dry season study (November through May, 1999-2000) and 761 in the rainy season (June through October 2000).

Trained field workers used structured questionnaires to gather data on episodes of diarrhea, and the recall period was the preceding week, as recommended by the World Health Organisation. The guardian (i.e., mother) also provided information on housing characteristics (e.g., water supply, sanitation, hygiene, and socioeconomic-related variables).

Data Management and Analysis

Both environmental and population data were entered twice and error corrected. IBM-compatible computers (486 processors) were used. For population data, the unit of analysis was the individual. Each child was allocated to one water quality category, which remained constant throughout the analysis; children exposed to different wells were allocated to the highest exposure.

An episode of diarrhea (health outcome) was defined as three or more loose stools in 24 hr. Potential confounding factors were included in the analysis; crowding (proxy of low socioeconomic strata) was incorporated as a continuous variable (three to six individuals per bedroom), and the odds ratio was interpreted as the excess of disease among children living in crowded dwellings, when compared with children without this factor (i.e., < 3 individuals). Special attention was paid to seasonal differences, and every independent variable showing significant association (Pearson chi-squared test) with diarrheal diseases was included in the final model. Statistical analysis was performed by using multiple logistic regression techniques.

Disease prevalence rates, odds ratios (OR), 95% confidence intervals (95% CI), and p-values were the measurements employed, by using STATA and SPSS. Both the environmental and health risk data were overlapped by using GIS, and MapInfo software was used.

The prevalence rates of diarrhea were 10.7% in the dry season and 11.8% in the rainy season. Crowding conditions were detected in more than half of the dwellings visited. Three-quarters of the children came from households with piped water supply inside their dwelling; for more than 60% of these, however, water supply failures (> 12 hr/day) were a common experience. Data showed that drinking water was usually stored in unprotected tanks and buckets (29-33%). When respondents were questioned about perceived characteristics of tap water, a third of them reported unpleasant taste; a similar proportion reported purchasing commercially bottled water, particularly during the dry season.

Wells N1 and N3 showed the presence of bacterial indicator (FC/100 mL), whereas wells N2, N6, and SL19 showed consistently negative results (i.e., absence of bacteria). Bivariate analysis showed no statistical association between the presence of bacterial indicators (FC/100 mL) in water samples and risk of enteric diseases (OR = 0.7 in the dry season and OR = 1.1 in the rainy season). As data illustrate, longer episodes of diarrhea were detected in the dry season than during the rainy season (OR, 3.7; 95% CI, 1.39-10.08), particularly among individuals between their first and second birthday (22.6%; OR, 2.1; 95% CI, 1.03-4.50).

The lowest prevalence of diarrhea was detected in older children (7.7% and 9.5%), although this observation was statistically significant only in the wet season study (OR, 0.5; 95% CI, 0.26-0.94). The prevalence of diarrhea was higher among children from households without piped water than among children in households with it (18.5% and 8.1%, respectively), and this association was detected only in the dry season (OR, 2.5; 95% CI, 1.58-4.18).

Children from households complaining of unpleasant taste of water had a higher risk than did children from households without such complaint (OR, 1.7; 95% CI, 1.04-2.70 in the dry season; OR, 1.5; 95% CI, 0.95-2.33 in the rainy season), whereas perceived colour of water was statistically significant only in the rainy season (OR, 1.9; 95% CI, 0.97-3.74) and not in the dry season, when a 2-fold risk was observed among children from dwellings without water for flushing the toilet (OR, 2.1; 95% CI, 1.32-3.41), as well as from dwellings where water was stored in unprotected buckets or cisterns (OR, 1.9; 95% CI,

1.04-3.42) and with no hand-washing habits (OR, 1.7; 95% CI, 0.95-3.13). Children from households whose members held cultural explanations of diarrhea had a higher risk of diarrhea than did children in households with beliefs regarding food and water pollution or hygiene (OR, 1.9; 95% CI, 1.05-3.70). Children who had recently consumed food sold by street vendors had a higher rate of diarrhea than did those who did not, although this association was found only during the wet season (OR, 1.7; 95% CI, 0.99-2.87). A health risk was also detected in children living in crowded dwellings (OR, 1.2; 95% CI, 1.03-1.50), although this association was observed only during the dry season.

Logistic regression analysis confirmed a lack of statistical association between the presence of fecal coliform bacteria in groundwater samples and health risk. The final analyses showed that the highest prevalence of diarrhea affected children 1-2 years old, whereas a decreasing risk was observed in older children; in the younger children, the difference was statistically significant in the dry season (OR, 2.1; 95% CI, 0.99-4.71), whereas the lower risk in older children was observed only during the wet season (OR, 0.5; 95% CI, 0.26-0.95).

Dry season data showed that children from households perceiving unpleasant taste of water had a higher risk than did children in households without such complaint (OR, 1.7; 95% CI, 0.97-2.92). In addition, a 2-fold risk was observed in children from households in which vegetables are usually washed only with tap water before consumption, compared with households using chlorine for disinfection or soap (OR, 2.2; 95% CI, 1.10-4.39).

In contrast, protective associations were observed among children from households with full-day water supply (OR, 0.5; 95% CI, 0.27-0.86) and having a flushing toilet (OR, 0.3; 95% CI, 0.16-0.67) and a shower inside the dwelling (OR, 0.4; 95% CI, 0.23-0.96). A similar pattern was detected for storing water in covered jars (OR, 0.3; 95% CI, 0.15-0.80). Rainy season data showed that children from households complaining of drinking water colour had a higher prevalence of diarrhea than did those in households without it (OR, 1.8; 95% CI, 0.93-3.67). Recent consumption (i.e., preceding week) of food from street vendors was also observed to represent a health risk (OR, 1.6; 95% CI, 0.98-2.87).

This investigation suggested an endemic pattern of enteric diseases, rather than a waterborne outbreak; the rates of diarrhea were not substantially different from the ones recently reported for Mexico as a whole. Equally important, perhaps, the water quality indicators used did not predict the health risk. It is necessary to emphasize, however, that groundwater is in jeopardy, and this could be the actual meaning of the presence of bacteria in water samples. Despite the lack of statistical association between groundwater quality and health risk, it is worth emphasizing that fecal pollution is finding its way to underground water sources. This observation may be different from official reports.

As expected, the rate of enteric disease was slightly higher in the wet season; more risk factors, however, were detected during the driest time of the year. These observations are not new, but reinforce the following points: The high prevalence of diarrhea detected in children within the first months after birth (rainy season) and a 2-fold risk in older children (dry season) may be suggesting different enteric syndromes with possible seasonal influences. The whole picture simply confirmed that housing deficiencies, hygienerelated behaviour (including food), water storage practices, and risk perceptions were all at play.

Interestingly, children from households complaining of unpleasant attributes of water (e.g., taste, colour), as well as those stating culturally influenced beliefs in disease etiology (e.g., evil eye), had a higher risk of diarrhea than did those without complaints and giving hygiene-related answers to questions regarding beliefs. Similar observations were reported by Whiteford studying the ethnoecology of water-based diseases in the Caribbean; our work reinforces the relevance of perception data in environmental research. Our study has limitations that must be taken into account.

First, eligible households were confined to less than 500 m around each well, assuming that children were not exposed to "distant" wells, which may or may not be the case.

Second, and equally important perhaps, groundwater samples were obtained before the water passed through the chlorination device, and therefore microbiologic results did not reflect directly the quality of water flowing through the distribution pipes and reaching the consumers; financial and logistical constraints prevented house-to-house tap sampling.

Third, as the data illustrate, more than a third of this population reported current consumption of commercially bottled water, whereas direct ingestion from the tap was seldom reported.

Methodologic shortcomings may also result from the cross-sectional study design, which does not prove cause and effect. It must be stated, however, that more than 75% of children involved were evaluated during both seasons (the rest were replaced from within the same compound). Furthermore, the involvement of a control group, the fact that water quality data were unknown to interviewers and respondents, the use of an operationally defined health outcome, and the procedures used to control for potentially confounding factors (e.g., socioeconomic status) during the analysis all reduced the chances of bias.

Final comments refer to future environmental health research: The challenge of detecting "waterborne" diseases represents a major issue, because methodologies are rather insensitive, labourious, expensive, and/or time-consuming. It must be emphasized that many intestinal infections may show few clinical symptoms, and outbreaks may not be detected. Previous exposure to enteric pathogens may alter a subject's clinical response, often reducing the severity of illness; when pathogens are endemic, much of the population may become immune. On the other hand, if pathogens are removed from drinking water ("zero" risk), the population may become increasingly susceptible.

The interpretation of the safety threshold (i.e., absence of FC/100 mL) requires further discussion, particularly in water reclamation scenarios. Earlier screening in our setting has detected Giardia intestinalis cysts and Cryptosporidium parvum oocysts in the water treatment plant effluents that comply with current quality guidelines (no detectable FC/100 mD. New studies are being conducted to address the question of risk from protozoal infection, and these results will be presented in future communications.

11

The Challenges of Sustainable Access to Safe Drinking Water

Water is a vital limited resource for human existence and the availability of adequate and safe water connects strongly with the sustainable development concept. Water plays a major role in enhancing economic growth, reducing poverty, attaining food security, and protecting ecosystems. Moreover, water is a critical component of public health, and failure to supply safe water will place a heavy burden on the entire population. Although the 20th century witnessed great wealth and advanced development in many areas of the world, billions of people are still striving to access the most basic human needs: food, shelter, safe drinking water, and sanitary systems. Closing this gap requires the availability of safe water for more than 1.1 billion people and sanitary systems for 2.4 billion people worldwide.

Water is subjected to biological and chemical contamination originating from many sources in the catchment area or during collection, storage, and distribution. Chemical and biological contaminants in water can cause many problems related to human health (diseases) or people's acceptance of water (a bad taste, odor, or colour). Water-related diseases are either infectious (caused by pathogens) or noninfectious (caused by toxic chemicals). While chemical pollutants in water can cause adverse health effects in the longterm, microbial pathogens can cause health effects within a short period of time.

Waterborne diseases are a great burden on both public health and the economy. Globally, four billion cases of diarrhea occur annually, of which 2.2 million are fatal (WHO & UNICEF, 2000). In developing countries, nearly 80%

of all diseases are linked to water and sanitation (WHO & UNICEF, 2004). Children bear the greatest health burden related to poor water and sanitation. Diarrhea is a major cause of death among children worldwide, causing death of one in five children and 15% of all deaths in children under age five.

Furthermore, diarrhea causes 54 million disability-adjusted-life-years (DALYs) each year, forming 3.7% of the total global burden of disease. Improving water and sanitation conditions is a critical tool to minimise diarrhea. Up to one-third of the cases can be minimised by an adequate access to water and sanitation and one-third can be reduced by washing with water and soap (WHO & UNICEF, 2000).

In Lebanon, water resources are progressively affected by natural and anthropogenic activities. It is estimated that 60%-70% of water resources are contaminated, with higher levels expected in the rural areas.

Moreover, rapid population growth in some areas was not associated with the required concurrent infrastructure development. As such, many areas are still suffering from water shortages. Improvements in sanitation and provision of water services continue to lag behind the needs of the population in almost all rural areas in Lebanon.

In the village of Zawtar El-Charkieh, our study area in southern Lebanon, drinking water sources are anticipated to be microbiologically contaminated due to the absence of sanitary methods to dispose of wastewater. As part of an ongoing partnership between the Faculty of Health Sciences at the American University of Beirut and the municipality of the village, an initial assessment study of the quality of drinking water was conducted in summer 2007. Results revealed that water is microbiologically contaminated in more than 60% of the collected samples. Also, people were concerned about the potential effects on the water supply system in Zawtar El-Charkieh and other villages in southern Lebanon after the July 2006 war.

People need to be reassured about water quality based on scientific data. Accordingly, the main objectives of our study were as follows:

- Assess the drinking water quality, cost, quantity, continuity, and coverage from the sources to the end users at the household level in Zawtar El-Charkieh.
- Identify the challenges of sustainable access to safe drinking water in Zawtar El-Charkieh village in order to determine the short-term management actions and long-term strategies to improve water quality.
- Inform decision makers and researchers on the challenges and approaches that could be applied in rural areas of other developing countries to ensure sustainable access to safe drinking water and facilitate in achieving the Millennium Development Goals (MDGs).

Description of the Study Area

The study area is located in Nabatiyeh province of southern Lebanon, 83 km away from the capital Beirut. It has 600 households and a total of 3,750 residents. It is located at an altitude of 340-490 m and covers an area of about 36.5 [km2]. The climate in this region is Mediterranean, characterised by seven hot months (May to October) with almost no precipitation and five cold months (November to April) with precipitation ranging between 200 and 1,400 mm. Households in the village are served by two piped water supply systems as follows.

- Al-Beer water supply network, which receives water from a well inside the village. Water is pumped out to the village to a sole storage tank of 450 [m^3] capacity from which it is distributed by gravity to the households.
- Nabeh Al-Tassi network, which receives water from a spring outside the village. Water is pumped and distributed directly to the households.

A complementary nonpiped water supply system, which derives water from a private well outside the village, delivers water by cisterns to the households. In addition, many people purchase bottled water to meet their demand.

Study Design and Sampling Strategy

The assessment encompassed all the drinking water sources in the village. A total of four samples were collected from each source in addition to the storage tank during the months of November and December 2007 and January and February 2008 to cover dry and wet seasons.

With regard to the terminal ends of the water distribution system, a total of 41 households were sampled once during the month of February 2008 and once during the month of April 2008.

The sample size was estimated using Equation 1. Based on the contamination prevalence of 60%, which was estimated from the preliminary assessment, the sample size was determined within 10% standard error and with 80% (z = 1.28) confidence interval.

n = [[(z)2] x p x q]/[d2] (1)

Where,

n = household sample size

z = confidence interval

p = probability of success

q = probability of failure

d = margin of error.

A total of 24 households at the terminal end of Al-Beer network and 17 households at the terminal end of Nabeh Al-Tassi network were chosen randomly based on the availability of taps connected directly to the water network. Samples were distributed among the seven zones in the village. Based on their corresponding approximate household number, samples were divided proportionately among the zones.

In order to track the source of contamination, drinking water sources and the storage tank were sampled on the same days the household samples were collected. Water coverage, quantity, continuity, and cost were assessed by interviewing key persons in the municipality and the water authority as well as the households that were sampled.

Analytical Procedures

All sampling equipment and containers were thoroughly washed in dilute (0.1%) low-phosphate detergent and tap water, then rinsed with dilute (1%) nitric acid solution (analytical grade) and deionized water. Except during processing, all samples were kept chilled (4[degrees]C) from the time of collection until analysis. Physico-chemical analyses were conducted at the American University of Beirut following standard and recommended methods (American Public Health Association [APHA], American Water Works Association [AWWA], & Water Environment Federation [WEF], 2005). Water temperature, pH, electrical conductivity (EC), and total dissolved solids (TDS) were measured on site using a Hach Model 44600 Conductivity/ TDS Metre. Turbidity was measured by using HACH 2100 P Turbidimetre. In addition, the dissolved oxygen (DO) was also measured on site using a membrane electrode Hach Model Sension 6.

Water samples were analysed for indicator parameters including total (TC) and fecal coliform (FC) bacteria, nitrate-nitrogen (N[O3-]_N), ammonia-nitrogen (N[H3]_N), chlorides ([Cl-]), and sulfates (S[O42-]). A spectrophotometre was used for N[H3]_N (Direct nesslerization; Standard Methods 4500 N[H3] C), N[O3-]_N (Cadmium reduction; Standard Methods 4500-N[O3-] E), and S[O42-] (Turbidimetry; Standard Methods 4500-S[O442-]E). Free residual chlorine was analysed by using HACH colorimetre. The bacteriological analyses were determined by the membrane filtration technique (Millipore). Titration method was used for the analysis of total hardness, calcium (Ca) hardness, Chloride ([Cl-]) and alkalinity. Total organic carbon (TOC) was measured by using PASTEL UV analyser.

The field quality assurance/quality control (QA/QC) component consisted of field blanks, equipment blanks, and duplicates. The QA/QC component of the analytical methods consisted of procedural blanks, lab duplicates, and reference samples (laboratory spikes and reference materials). The results from duplicate analyses were averaged for use in this study.

Water Quality

All samples collected from Al-Beer well were microbiologically contaminated with levels exceeding the recommend World Health Organisation (WHO) and United States Environmental Protection Agency (U.S. EPA) value of 0 CFUs/ 100 mL for total and fecal coliform bacteria. This manifests a relationship between the quality of water and the land use and human activities within the catchment area. The levels of total coliform bacteria ranged between 27 and 320 CFUs/100 mL. Likewise, fecal coliform bacteria (0 to 71 CFUs/100 mL) were detected in almost all samples with the exception of the sample collected at the end of the wet season. Predictably, the water samples collected from the storage tank were all microbiologically contaminated with the exception of the sample collected at the end of the wet season, which was free from fecal coliform bacteria. This result may be attributed to the contamination of the water source, the infrequent cleaning schedule of the tank, and the fact that no disinfection method is being practiced.

By contrast, almost all the samples collected from Nabeh Al-Tassi spring were free from fecal and total coliform bacteria, which may be attributed to the fact that water is being chlorinated. The free residual chlorine ranged between 0.19 and 0.25 mg/L. Similar to the Al-Beer well, poor water quality conditions were observed at the private well, which delivers water by cisterns to the households. The samples collected from this complementary nonpiped water supply system (private well) were all microbiologically contaminated, with the exception of the sample collected at the end of the wet season that was free from fecal coliform bacteria.

All water samples collected at the end of the wet season from the households connected to the Al-Beer network were contaminated with total coliform bacteria. Although the Al-Beer well and the storage tank were free from fecal coliform bacteria at the end of the wet season, about 42% of the samples collected at the household level were contaminated with fecal coliform bacteria. This may be attributed to improper sewage management. About 29% of the samples collected from households connected to Nabeh Al-Tassi network revealed levels of fecal and total coliform bacteria, which may be attributed to a reduction in free residual chlorine. The concentration of residual chlorine decreased from 0.19 mg/L at the source to 0.02 mg/L at the distal ends.

Physical and Chemical Analysis

Physical and chemical water analyses were conducted on samples taken from the three water sources: Al-Beer well, Nabeh Al-Tassi spring, and the private well. As a result of seasonal variation, the temperature values decreased between the dry and the wet season for the three water sources. Slight variations were exhibited in the levels of pH between the dry and the wet season with values ranging between 6.4 and 7.4. While the levels of turbidity decreased between the dry and the wet season for Al-Beer well and Nabeh Al-Tassi spring, higher levels were observed for the private well. This result may be attributed to a dilution factor in the case of the former sources and maintenance operations during the sampling period for the latter. The concentrations of DO slightly increased during the wet season, which may be attributed to a decrease in temperature.

The values of TOC exhibited a slight increase between the dry and the wet season for Al-Beer and the private well, which may be a result of wastewater contamination. Due to the dilution factor, the levels of TDS exhibited a slight decrease between the dry and the wet season for all sources. Hydroxide alkalinity exhibited a value of zero in all samples. As such, total alkalinity is attributed to bicarbonate, which decreased between the dry and the wet season as a result of dilution factor. Total hardness in all samples exhibited values greater than the total alkalinity. Accordingly, total hardness is attributed to both carbonate hardness and noncarbonate hardness. All sources exhibited a decrease in the levels of total hardness, calcium, and magnesium between the dry and the wet season, which may also be attributed to dilution factor.

Generally, the concentrations of N[O_3-]_N and N[H3]_N exhibited a decrease between the dry and the wet season as a consequence of dilution. With the exception of Nabeh AlTassi spring, the concentrations of S[O42-] and [Cl-] exhibited an increase between the dry and the wet season. Moreover, S[O42-] and [Cl-] to TDS ratio increased between the dry and the wet season, which may be ascribed to the build up of their concentration as a result of sewage contamination. Compared to the WHO and U.S. EPA recommended values for drinking water, the measured concentrations of all physical and chemical indicators were within these limits.

Water Coverage, Quantity, Continuity, and Cost

Of the 600 total households in the village, 444 are connected to Al-Beer network, 150 are connected to Nabeh Al-Tassi network, and six are not connected to the networks. About 61% of the houses are legally connected (subscribed) with a gauge system. The remaining houses are illegally connected. Unaccounted-for water (physical and commercial water loss) in the networks is expected to be as high as the Middle East and North Africa region, with a value of 52% due to aging of the networks. Given a total population of 3,750 residents and a total number of 600 households, the average number of persons per household is 6.25. A gauge system that supplies one cubic metre per day is being utilised in the households connected to Al-Beer network. While pumping hours in the case of the Al-Beer network may reach up to 16 hours/day for two to three days per week year-round, Nabeh Al-Tassi network provides water on a 24-hour basis for two (summer) to four (winter) days per week. Accordingly, the estimated per capita water quantity that people receive is

42 L/capita/day in summer and 64 L/capita/day in winter. This is mainly attributed to the limited pump capacity and electricity rationing in the case of Al-Beer network and the over demand that exceeds the supply of Nabeh El-Tassi network.

Both networks in the village work intermittently. People are used to using storage tanks to store water during pumping hours. In addition, they purchase water, particularly in the summer, to meet their needs. The monthly family income ranges between $495 and $550 U.S. The average monthly expenditure that each household pays for the public water subscription is 20,144 Lebanese Lira (LL) ($13.50 U.S.) and the complementary water (bottled and cistern water) is 41,928 LL ($28 U.S.). As such, total cost of water per month is equivalent to $41.5 U.S., constituting about 8.4% of the average total family income.

Water contamination, primarily bacteriological verified by the presence of total and fecal coliform bacteria, may be attributed to the improper management of wastewater. Due to the absence of a sewage network, unsanitary pits are being utilised to dispose wastewater. Contamination of the source, absence of any disinfection method or insufficient dose, poor maintenance operations, and aging of the networks are significant factors contributing to water contamination during storage and distribution process. The fragile political stability, restrictions, and insecurity in southern Lebanon for almost two decades have led to a lack of water development plans. This situation is still ongoing and will probably not come to an end in the near future. In addition to the resources and educational challenges, political disputes are major stumbling blocks to achieving safe drinking water conditions. Needless to say, economic and political considerations in spite of everything tend to manipulate water policies and strategies.

Although the results revealed that chemical quality of drinking water meets the WHO and U.S. EPA standards, seasonal variation in the concentrations of various indicators demonstrates the influence of anthropogenic activities and land use on water quality. In addition to improper sewage disposal, contaminants may originate from runoff from fields where crops are irrigated and the use of inorganic fertilizers (usually as ammonium nitrate) and animal manure is substantial. Thus, the qualitative and quantitative improvement of agriculture productivity resulting from agrochemical use is counter-balanced by the detrimental consequences on human health, environment, and hydrologic systems.

Since a public water supply is cheaper than private sources, almost all the households in the village are subscribed to the public network. Yet, the quantity of water is far below the estimated water consumption of 190 L/capita/day in Lebanon. The quantity of water provided to the community is far below neighbouring countries in the region such as Palestine, Jordan, and Egypt where the values range between 90, 95, and 183 L/capita/day, respectively. This shortage may be ascribed to a large amount of water that has not been accounted for due to physical and commercial water loss, aging of the networks with minimum maintenance operations, and illegal connections. The interruption in the water supply also may partially contribute to the acute water shortage in the village. Hence, local people seek other sources to meet their water needs with average monthly expenditure reaching 8.4% of the total family income. This value exceeds the recommended international value of less than 2%. Therefore, many people may be forced to seek lower-priced water, which in turn may be of lower quality, or they may minimise their hygienic activities. This inflicts a significant threat on the health of the local community due to waterborne diseases.

While there are many impediments and challenges towards achieving safe, affordable, and sustainable access to drinking water, these can be overcome by suitable planning and proper measures. Although it may seem costly, rehabilitation of both networks will be cost-effective in the long-term by decreasing water loss and improving water quality.

Yet, new sources of water as well as community participation in conservation practices are essential to achieve sustainability and reduce water shortages. Given the limitations of external and domestic financial resources in developing countries, it will be necessary to develop new innovative financial schemes that could include public-private partnership in financing. A high priority should also be given to preventive actions, such as protection of the catchment areas of water sources.

An optimised operation strategy for wastewater management in rural areas should be developed. Such a strategy should be based on an integrated assessment of the local technical, environmental, economic, and social aspects. Determining the nonpoint sources of pollution would be imperative in protecting the sources of water. Since nonpoint sources of pollution are diverse and nonspecific, however, control and prevention measures through the application of conservation principles that are ecologically sound should be developed. As such, environmentally sound national agricultural policies should be introduced in addition to controlled and optimal use of agrochemicals.

Considering that drinking water is not effectively monitored in these areas with low population density, several problems may arise. Under conditions of poverty, unemployment, and political instability, few incentives or sufficient capital are present to employ environmental monitoring programs. Environmental issues do not always command a high priority in light of the severe social, political, and economic problems that face most developing countries. If sustainability is to be achieved, a clear institutional and regulatory framework with quantitative objectives and standards for decision making; a system that provides reliable information on the availability, use, and quality of water; as well as adequate funding for management and operation are crucial. Regular monitoring of residual chlorine concentration as well as fecal and total coliform bacteria in the water distribution system should be carried out systematically by the local water authorities or municipalities.

In the short-term, simple but effective measures can be taken to ensure the protection of water resources and prevent the buildup of contaminants. Increasing the residual chlorine dose is essential to reducing microorganisms in the distribution network. The presence of residual chlorine improves sanitary conditions, suppresses the regrowth of microorganisms, and prevents the formation of biofilms on the internal surface of distribution pipelines. Training programs for employees of the municipalities and water authorities are essential for the proper operation and maintenance of equipment and facilities, including monitoring of water quantity and quality. Educational material with illustrations and understandable information on the quality of drinking water, risk of water contamination, and how to protect water sources from contamination and keep water clean and safe is very crucial.

This would empower the local community to protect themselves against waterborne pathogens and consequently reduce waterborne diseases. Promoting public awareness of the long-term environmental and economic benefits of water conservation and identifying habits and efficient water devices influence community behaviour.

The majority of the challenges of sustainable access to safe drinking water in Zawtar El-Charkieh are also common to other small villages in Lebanon as well as many rural areas in developing countries. These communities often suffer from an inadequacy of services related to health care, nutrition, and public education in addition to inadequate infrastructure and very low income levels. Therefore, governments in developing countries have more pressing needs than water management. Environmental issues are frequently low on the list of priorities. Public awareness relating to the extent of adverse health impacts as a result of unsafe drinking water and improper sanitation is minimal in these countries.

Meeting the goal of sustainability and safety of drinking water sources and consequently the MDG goal for water is a major problem in Zawtar El-Charkieh and in many rural areas in developing countries. In light of the aforementioned, environmental education as well as public awareness and community empowerment should be given a high priority. Providing local people with access to resources, education, and information necessary to influence environmental issues that affect them is a necessity. To ensure public health protection and if sustainability is to be achieved, however, establishing a comprehensive drinking water system that integrates water supply, quality, and management as well as a clear institutional and regulatory framework with quantitative objectives and standards for decision making associated with adequate funding for management and operation are crucial.

Holy and Wholesome

Among the blessings of working in the multilateral, international and interdisciplinary context of the World Council of Churches (WCC) are the many different dialogue partners who are able to challenge, broaden and mutually enrich one another. One such dialogue has been between a Canadian Anglican liturgical theologian and a German Lutheran eco-justice activist. Over a series of months Maike Gorsboth, coordinator of the Ecumenical Water Network, and John Gibaut, director of the Commission on Faith and Order, have engaged in a challenging and enlightening dialogue on the issue of water. We have already collaborated in the successful preparation of texts for the 2010 Seven Weeks for Water, which address the ecological crisis of water and its accessibility from different liturgical perspectives.

We are aware that to some, if not many, the Ecumenical Water Network and Faith and Order are unusual dialogue partners within the WCC. While the Ecumenical Water Network focuses on mobilising concrete ecumenical activities to promote good stewardship and just access to water, Faith and Order engages in long-term multilateral theological reflection which seeks to resolve church-dividing issues, many of them relating to liturgical and sacramental questions, including water baptism. In this chapter we would like to share some of the learning we have gleaned from one another and our conversations, including the challenges posed by eco-justice advocates to the churches in their liturgical use of water and the challenges that the liturgical tradition poses in the context of the water crisis. In the end, we affirm that liturgical and activist perspectives belong together and are mutually enriching.

The Global Water Crisis

Today, humanity faces the overuse and pollution of water to an unprecedented degree, threatening ecosystems and the health and livelihoods of billions, particularly of the most vulnerable. Economic growth, population growth, changing life-styles and urbanisation are increasing the demand and therefore the pressure on our world's water resources. Climate change is an important additional factor, aggravating water scarcity and leading to more extreme and less predictable weather patterns that affect how much water is available, where and when.

Those who suffer most from this situation are the world's poorest. Of the almost 1 billion who do not have adequate access to safe drinking water today, almost all live in developing countries and two out of three are surviving on less than $2 a day. Many of them rely on dirty water from open rivers, lakes and wells, or spend several hours collecting water every day. Others spend large parts of their restricted income to buy water.

Moreover, half of the more than 2 billion people who do not have access to adequate sanitation belong to the world's poorest households. The unsafe collection and disposal of human excreta is closely linked to the contamination of food and water and the spread of water-related diseases that kill 2 to 3 million people every year. WaterAid estimates that inadequate water and sanitation may be the underlying cause of more than a quarter of the 8.8 million deaths among children under five each year.

It is also the poorest who are hardest hit by increasingly unpredictable and more extreme droughts and floods. Seventy percent of the world's poorest live in rural settings where they depend on rearing livestock and on mainly rain-fed agriculture to feed themselves and their families and generate some income. Already today about 700 million people live in areas affected by water stress; by 2025 this figure could be up to 3 billion.

For many governments and communities, it is becoming more and more difficult to satisfy the different demands and needs beyond drinking water: water to ensure food security, water to supply rapidly growing cities and industries, water for ecosystems. Ecosystems collapse as the natural water cycle is altered and the dumping of our waste overwhelms nature's capacity of purification. In many places the very "bloodstream of the biosphere" is at risk of being clogged.

While the exact nature of water problems and their possible solutions differ from one situation to another, it still seems appropriate to talk of a "global crisis" for a number of reasons. First, even the promotion and implementation of local solutions require global solidarity and analysis. Second, there are many cases where water is depleted and polluted in the production of food and goods that are sold and consumed far away from the local populations suffering the consequences. Similarly, climate change is not always caused where its impact is most felt. Third, some of the underlying causes of the crisis are global in nature: the often rather single-minded pursuit of economic development, the inefficient and wasteful use of water resources, inadequate management and regulation, and all too often the insufficient prioritisation and protection of the needs of the poorest.

Faith and Water

For us as inhabitants of our planet, water is an issue that concerns all Christians, raising the topics of just stewardship of the gift of God's creation, just access to water and just usage of water. This idea was expressed by the Ninth Assembly of the WCC at Porto Alegre in 2006:

Water is a symbol of life. The Bible affirms water as the cradle of life, an expression of God's grace in perpetuity for the whole of creation (Gen 2:5ff). It is a basic condition for all life on Earth (Gen 1:2ff.) and is to be preserved and shared for the benefit of all creatures and the wider creation. Water is the source of health and well-being and requires responsible action from us human beings, as partners and priests of Creation. As churches, we are called to participate in the mission of God to bring about a new creation where life in abundance is assured to all (John 10:10; Amos 5:24). It is therefore right to speak out and to act when the life-giving water is pervasively and systematically under threat. The deep Christian concern for water is shaped by the biblical narrative beginning with the waters of creation and ending in the waters of the new creation, the river of the waters of life coursing through the New Jerusalem, bringing healing to the nations. The narratives of life and death and new life in the Hebrew and Christian scriptures are often described in the context of water, from the Flood and the Exodus to the narrative of the baptism of Jesus, the stories of events around the Sea of Galilee in the gospels, and Paul's preaching of the Gospel across the waters of the Mediterranean. The ordinary, everyday experience of water as healing, cleansing, quenching, life-giving, as well as death-dealing, and the bountiful biblical instances of water as part of God's healing action, are brought together within the context of the liturgy, in particular in baptism. Paul makes this connection clear in the letter to the Romans, where Christian participation in the death and resurrection of Jesus takes place in the experience of being baptized—literally "plunged under"—and raised up in the waters of baptism.

Water plays a pivotal role in the life of the church through baptism and consequent liturgical acts, such as the blessing of people and things through blessed water, the celebration of the baptism of Jesus in the waters of the river Jordan, the place of water in the Byzantine celebration of the Theophany, and other liturgical acts. The theme of water appears in the course of the lectionary or assigned biblical readings throughout the liturgical year. And, for many traditions, water is liturgically blessed for use in baptism. Given the pivotal role of water in the liturgical tradition, it is curious that sacramental and liturgical theologians on the one hand, and eco- or environmental activists and theologians on the other, are not in frequent alliance with one another for the enriching of the experience of water for the Christian community at prayer, and for strengthening the prophetic and advocacy role of the church around a just and faithful stewardship of water as a part of God's creation.

Liturgical Theology

An axiom of contemporary liturgical theologians is derived from a fifth-century theologian, Prosper of Aquitaine: lex orandi, lex credendi. The Latin tag in ancient and contemporary literature states that the "law of praying" establishes or constitutes the "law of believing". In other words, how we pray shapes what we believe. Here, liturgy is understood as an event or experience, from which theological reflection follows; it posits that the experience of the encounter with God precedes its theological reflection. Many add a third factor, namely the "law" of acting or living, forming a neat triad: lex orandi, lex credendi, lex vivendi. Accordingly, if experience gives rise to reflection or meaning, and horizons of meaning give rise to concrete actions, then the experience of God in liturgy shapes Christian believing; this in turn shapes Christian action.

The axiom refuses to allow experience, belief and praxis to drift apart. Few would want to limit this dynamic interaction to liturgical prayer alone, but, as Aidan Kavanagh has poetically asserted, For if theology as a whole is critical reflection upon the communion between God and our race, the peculiarly graced representative and servant of cosmic order created by God and restored in Christ, the scrutiny of the precise point at which this communion is most overtly celebrated by us under God's judgment and in God's presence would seem to be crucial to the whole enterprise.

The intimate relationship between liturgy and theology is borne out through study of the development of doctrine, as well as in pastoral practice. For instance, the Nicene-Constantinopolitan Creed, while developed by two ecumenical councils, finds its historic origins in the earlier baptismal creeds recited during the liturgical celebration of baptism.

A simple example of pastoral life that illustrates the same development in individual Christians is found in the impact of Christmas carols and hymns on how we understand the incarnation. At times, insights into the validity of lex orandi, lex credendi are derived from negative experiences of corporate liturgical prayer; when liturgy is consistently sloppy or dull, it comes as no surprise when personal faith and piety falls away. An important insight about the power of liturgy in shaping belief and action arises from feminist critiques of liturgy, which point out that when exclusively male imagery is used for God or the human person, maleness becomes normative, with a decreased accent on the place and insights of women.

It is chastening to remember that 20th-century apartheid in South Africa began as liturgical division around access to the celebration of the Lord's supper and holy communion in the 19th century. In short, what and how the Christian community prays has profound and often serious consequences for what it believes and how it behaves. We have wondered whether part of the reason that the sacramental and liturgical use of water has played such a weak role in Christian discourse on the environmental and accessibility issues around of water lies in the all too frequent impoverished use of water in baptism in particular and in the liturgical life of the church in general.

Liturgical Praxis and the Water Crisis

In the understanding of liturgical theology, liturgy is an event rather than a set of prayer texts, so how the Christian community uses water is as important as what it says about it in liturgical texts. The earliest Christian practice of immersing children and adult candidates for baptism in water continues in some traditions, notably the Orthodox traditions. It was restored by Anabaptist and other traditions in the 16th century and recovered more widely in some of the Western liturgical traditions in the 20th century. Nevertheless, for most Western Christians, baptism by sprinkling or pouring a small amount of water on the forehead is still normative. The gradual move from immersion to sprinkling in the Western traditions from late antiquity through the medieval church is due to a variety of factors, such as the normative candidacy of infants, the corresponding privatizing of baptismal celebration and the spread of the centres of Western Christianity to more northern climates. As baptism became more and more clinical and baptismal fonts smaller and smaller, the minimalist instrumentality of water came to dominate. As such small amounts of water could no longer correspond to the reality of water in daily life, or evoke the deep biblical narratives of water, the rite of baptism could no longer "exegete" itself, and became overly dependent on words and explanation, and less rooted in the inherent and rich experiential symbolism of water.

An important question is the extent to which the diminished experience of water in the church's central act of Christian initiation, and its disconnection from water in daily life, may gave rise to a diminished sense of the holiness of water, with a corresponding absence of Christian responsibility for the health and wholeness of water. It is worth asking to what extent the theological devaluing of water, with its roots in liturgical praxis, led Western society to treat water as a commodity to be used and abused, rather than as a holy gift. While the roots of the current crisis are of course more complex, there is a link between the beliefs about water from the major Western religious tradition, Christianity, and attitudes that have misused water for centuries.

The sources of the rise in Christian consciousness around the environment, together with the emergence of eco-theology evident in the works of theologians such as Thomas Berry, Larry Rasmussen, Leonardo Boff and Rosemary Ruether, are rich and varied, with strong roots in the biblical witness. It is not surprising that the liturgy, understood as event or action, is seldom cited. Through the liturgical movement, however, there has been an insistence on the recovery of the fullest use of all the sacramental symbols, including eucharistic bread and wine, oil and touch.

The recovery of baptismal pools, immersion baptism with larger fonts, as well as the baptism of adult candidates has been part of a wider liturgical reappropriation of the central place of water. Locating the baptismal pool or font to a central place in the church building, often accompanied by flowing water, begins to match the use of water for bathing or swimming, for healing and the quenching of thirst; it begins to evoke the biblical imagery of water.

Such richer experiences of water give rise to a more fulsome theological understanding; it is at this point, where praxis and reflection meet on a different level, that liturgical theology and environmental theology and action begin to meet. Here, environmental concerns challenge the churches to live up to the theological vision of water inherent in baptism. It is here that the holiness of water is reflected in prayers of blessing or thanksgiving over it and its use in Christian initiation, and that the identification of gift and holiness in all water is authentically proclaimed.

Our conversations about the points of convergence between liturgical praxis and the environmental crisis around water have arisen from two "case studies" in particular arising from the articles prepared internationally and ecumenically for the 2010 Seven Weeks of Water. The first touches on the water crisis in terms of pollution, while the second touches on the issue of access to water. Both point to the radical discontinuity between liturgical understandings of water and the realities created by the global water crisis.

Holy Water—Please Do Not Drink!

From Elias Abramides, an Orthodox theologian from Argentina, comes an account of the Feast of the Theophany. The blessing of water occurs in many different liturgical traditions, usually for the celebration of baptism, and, by extension, for other blessings of water. For Eastern Christians, particular blessing of waters takes place on the Feast of the Theophany, the Eastern equivalent of the Western celebration of the Epiphany on 6 January. Whereas in the West, the "epiphany" or manifestation of Christ is to the Gentiles in the visitation of the Magi to the Christ Child, for Eastern Christianity the "Theophany" or "appearance or manifestation of God to the world" focuses on the Baptism of Christ in the River Jordan by John.

The Orthodox celebration of Theophany includes a blessing of water using a rich and powerful variety of biblical images and liturgical actions. If possible, the blessing of water takes place outside the church building, by living water such as a river, stream, lake or sea. The blessing includes dipping lighted candles into the water, breathing over the water, and dipping a cross in the water, then plunging the cross into the living waters of the river, stream, lake or sea. The faithful and their homes are blessed by being sprinkled with the holy water, as well as by drinking it.

As celebrated in settings such as the monastery Valamo in Finland, this ritual can bring the beauty of creation before the eyes of worshippers, and perhaps also some of its terror: the waters of baptism as a symbol of both death and rebirth, when you take the traditional bath in the ice-cold waters, as is customary in some countries in the northern hemisphere.

This year, a few days after Theophany, newspapers and internet sites around the world carried a news item with the headline "200 people hospitalised after drinking dirty holy water". It was reported that hundreds of people in the east Siberian city of Irkutsk in Russia had fallen ill after drinking from unprotected water sources on the date of the Theophany, in the belief that any water collected on Theophany was blessed and had curative powers. Apparently the cause was not water that had been blessed during the ceremony, but water from another, unprotected and unsafe source of water. However, it is worth thinking about the possible implications for a moment.

Some reactions to this news item expressed in comments on the internet showed little concealed ridicule and Schadenfreude, though these were mainly, but not exclusively, restricted to anti-religious platforms. Many remarked on how naive it was of people to think that blessing water would at the same time purify it. One commented that "maybe they should have added holy chlorine".

Leaving aside the apparent desire to ridicule "overly credulous ... believers", these reactions are interesting in our context as they reflect something else: the understanding that it is "common sense" to regard any open water source as likely to be harmful. Admittedly, even in Jesus' time it would have been common sense to be aware that there were both safe and unsafe sources of water.

Today, however, with up to 12,000 cubic kilometres of severely polluted fresh water—that is, one-eighth of all global resources—the scale of the current crisis is one that would have been unimaginable in pre-industrial times, is For instance, in China, more than half of all surface water is too contaminated to be used for human consumption. In Latin America, less than 14 percent of human waste is being treated; the untreated water ends up either directly or indirectly polluting rivers, lakes and groundwater. The chapter on the worshippers who fell ill from drinking contaminated water on the feast of Theophany mentioned that generally, tap water in Russia was undrinkable.

We asked Elias Abramides what it means if the Feast of Theophany is celebrated over a polluted body of water. His answer was clear: "It does not matter; it is still Holy Water." But holy polluted water? Shouldn't something in us rebel at the very thought? Although not an water is liturgically "blessed", all water is inherently holy. Its quality of "normal", "daily" water lies at the roots of its biblical and symbolic meaning and the churches' liturgical use of (holy) water, which reflects this meaning. Here, the liturgical blessing of water, revealing its inherent holiness as a gift of God, stands in radical contrast to the human misuse of water which, in a dangerously sacrilegious fashion, renders the holy unwholesome. The discontinuity between blessed-yet-also-contaminated water stands as a judgment against human abuse of God's creation. As such, polluted water blessed at the Feast of the Theophany becomes an intolerable liturgical scandal that impels Christians towards a just stewardship of God's creation.

Come to the Waters—Access to Water and Inequality

In the Great Commission at the end of Matthew's gospel, Jesus says to his disciples, "Go therefore and make disciples of all the nations, baptizing them in the name of the Father and of the Son and of the Holy Spirit." The universal call to baptism contains an assumption that there will be universal access to water. The global water crisis raises serious questions about the accessibility of water for all and, hence, of universal access to the waters of baptism.

From Lucy Wambui, a Presbyterian from Kenya, comes an account of baptism in an African context. She recounts a recent celebration of about 80 adults from the Maasai community of Kajiado (a semi-arid area in Kenya). They were baptized in a constructed pool, which she describes as a "deeper" version of a trough. Because the rivers are dry, water for the baptism had to be brought in from elsewhere, and was bought at the price of 2,000 Kenyan shillings (about 25 Swiss francs or US$35). In an area of Kenya where people and livestock are dying due to lack of food and water, this is a great deal of money, and represents a remarkable commitment to the liturgical practice of immersion baptism. Here, the people opted for the water of baptism, rather than buying food or drinking water.

After the celebration of baptism was over, the members of the congregation scrambled to collect the remaining baptismal water in order to feed their livestock. As Lucy points out, this story "represents the reality for many communities." It speaks of the poverty and lack of water and reflects the great inequality and injustice in access to water, as well as the potential for conflict. To some, the use of holy water for livestock could be understood as a sacrilege; to us, however, the sacrilege is the lack of access to water, including the waters of baptism, at least for those who cannot afford the high costs of water.

As highlighted before, and as illustrated by Lucy's story, the poorest suffer most from the shortage of water: part of that suffering is that they pay the most for water. Most of the urban poor in developing countries rely on usually unregulated private water vendors; the result is that they pay many more times the price for water than those who are connected to a publicly run or regulated water supply network. For the poor, living with the consequences of an unchecked commercialisation of water is a bitter reality. This point is something to take into account when the use of the market logic is promoted as a way to use and allocate water more effectively. What are the effects for the economically and politically less empowered? How can we ensure that water both is effectively used and contributes to economic 'development', yet ensure access to water as a human right, water for sustainable livelihoods, for creation, as well for sustaining cultural and religious traditions? In this context, Steve de Gruchy called for an "olive agenda", recognising the interconnectedness and aligning the brown agenda of (economic) liberation from poverty with the (ecological) green agenda of caring for creation.

It is profoundly disturbing to see in this story how the reality of water affects people's ability to fully live their liturgical practices, how the crisis of water intrudes on people's experience of God in prayer and worship. For some, it might be surprising that a poor and water-deprived community would opt to pay so much money to be able to have an immersion baptism. Lucy Wambui concludes that "the people who thirst most for physical water thirst even more for baptismal water."

The celebration of baptism promises spiritual salvation and the quenching of our spiritual thirst with the "living water"; it introduces the one who is baptized into the Christian community. Baptism is a universal promise that makes no difference between the poor and the rich. There is a strong discontinuity between the symbolism of this liturgical practice and the thirst and struggle for water in the real physical world. As the church offers the waters of baptism freely to everybody who wishes to drink of the living water and never thirst again, should not the same celebration remind us of the physical thirst and the discrimination from which so many suffer?

The insights of liturgical theology propose that how we pray shapes how we believe, and that how we believe shapes how we live and act in the world. Regarding the place of water in the prayer of the church, especially in the celebration of baptism, the experience in many traditions and communities, especially within Western Christianity, has been weak and minimalist. As such, water in its fullness cannot bear or reveal by its own nature its inherent meaning as something both life-giving and death-dealing, which quenches, heals, cleans, restores and kills. Such a self-exegesis is suggested in Faith and Order's text on the polyvalent nature of baptism in Baptism, Eucharist and Ministry:

Baptism is the sign of new life through Jesus Christ. It unites the one baptized with Christ and with his people. The New Testament scriptures and the liturgy of the Church unfold the meaning of baptism in various images which express the riches of Christ and the gifts of his salvation. These images are sometimes linked with the symbolic uses of water in the Old Testament. Baptism is participation in Christ's death and resurrection (Rom. 6:3-5; Col.

2:12); a washing away of sin (I Cot. 6:11); a new birth (John 3:5); an enlightenment by Christ (Eph. 5:14); a reclothing in Christ (Gal. 3:27); a renewal by the Spirit (Titus 3:5); the experience of salvation from the flood (I Peter 3:20-21); an exodus from bondage (I Cor. 10:1-2) and a liberation into a new humanity in which barriers of division whether of sex or race or social status are transcended (Gal. 3:27-28; I Cor. 12:13). The images are many but the reality is one.

The diminished experience of water in the church's central act of Christian initiation, disconnected from water in daily life, reflects a diminished theological appreciation of the holiness of all water, with a diminished sense of Christian responsibility for the health and wholeness of water. While again we underline that the roots of the current crisis are more complex, there is interrelatedness between the beliefs about water from the major Western religious tradition, Christianity, and attitudes that have misused water for centuries. The weakened lexorandi contributes to weak leges of both theology and action around environmental protection of water and just accessibility. Here, the context of the water crisis and the insights of eco-justice activists call for a richer and more holistic use of water by the church at prayer that allows water to speak for itself if it is to be considered holy by usage and blessing. At the same time, the two examples we noted of a rich liturgical use of water—the Orthodox Feast of the Theophany and a Reformed celebration of baptism incorporating full immersion into water—reflect a serious discontinuity between practice and theology of water and the lex vivendi. In these instances, the discontinuity raises liturgical questions of sacrilege. In the first, liturgical practice stands in judgment over any human practice that pollutes water and renders it undrinkable, if when theologically it is considered holy. In the second, liturgical practice stands in judgment over human practices that deny or limit access to water. Here, the liturgical tradition of the Church convokes Christians to seek and advocate for a just and ecological coherence between what it says in word and deed about water and what it says about the effects of the water crisis. We end our reflection with the hope that we may believe what we pray and live what we believe:

> *"In baptism we use your gift of water,*
> *which you have made a rich symbol*
> *of the grace you give us in this sacrament.*
> *At the very dawn of creation*
> *your Spirit breathed on the waters,*
> *making them the wellspring of all holiness."*

Bibliography

Anderson, Mary P. and William W. Woessner: *Applied Groundwater Modeling: Simulation of Flow and Advective Transport.* Academic Press, San Diego, 1992.

Asaithambi, S. : *Economics of Ground Water Management in India*, Abhijeet, Delhi, 2008.

Bhakar, S.R. : *Ground Water Hydrology : Theory and Practice*, Agrotech, Delhi, 2009.

Bhave, P.R. and R. Gupta: *Analysis of Water Distribution Networks*, Narosa, Delhi, 2011.

Biswas Asit K. : *Integrated Water Resources Management in South and South-East Asia*, , Oxford University Press, Delhi, 2001.

Bourne, Peter G.: *Water and Sanitation: Economic and Sociological Perspectives* Academic Press. Orlando. 1984.

Brooks, D.: *Water: Local-Level Management*, Ottawa, International Development Research Centre, 2002.

Cullis, A. : *Rainwater Harvesting: The Collection of Rainfall and Runoff in Rural Areas,* London, U.K.: IT Publications, 1986.

Erik Nissen-Peterson : *Rainwater Catchment Systems*, UK: Intermediate Technology Publications, 1999.

Feder, Paul A.: *A Guidebook to Groundwater: Resources and Education Opportunities in the Great Lakes Region*, Great Lakes Commission, Ann Arbor, 1993.

Fetter, Charles Willard: *Applied Hydrogeology*, Merrill Publishing Company, Columbus, OH. 1988.

Geraghty, James J.: *Water Atlas of the United States*, Port Washington, NY, 1973.

Gischler, Christiaan E.: *Water Resources in the Arab Middle East and North Africa*, Cambridge, Eng, Middle East & North African Studies Press, 1979.

Gurjar, Ram Kumar : *Geography of Water Resources*, Rawat, Delhi, 2008.

Hemenway, Toby : *Gaia's Garden: A Guide to Home-Scale Permaculture*, Vermont: Chelsea Green Publishing Company, 2000.

Jat, M.L. and S.R. Bhakar: *Ground Water Hydrology : Theory and Practice*, Agrotech, Delhi, 2009.

Kanmony, J. Cyril : *Drinking Water Management : Problems and Prospects*, Mittal Pub, Delhi, 2010.

LeRoy, K. : *Sustaining Water Washington*, D.C.: Population Services International, New York, 1993.

McGarry, M.G.: *Matching Water Supply Technology to the Needs and Resources of Developing Countries,* United Nations. New York. 1987

Meenu Bhatnagar: *Groundwater Management : Sustainable Approaches*, Icfai Books, Delhi, 2012.

Morris, Mary E.: *Water Scarcity and Security concerns in the Middle East*, Abu Dhabi, Emirates Center for Strategic Studies and Research, 1998.

Narwani, G.S. : *Community Water Management*, Rawat, Delhi, 2005.

Nielsen, David M.: *Practical Handbook of Groundwater Monitoring*. Lewis Publishers, Chelsea, 1991.

Palanisami, K : *Groundwater Management and Policies*, Macmillan Publishers India, Delhi, 2008.

Premjit Sharma: *Agricultural Drainage and Water Quality*, Gene Tech Books, Delhi, 2007.

Singh, A.K.: *Environment and Water Resources Management*, Adhyayan, Delhi, 2006.

Thomas, Harold E.: *The Conservation of Groundwater: A Survey of the Present Groundwater Situation in the U.S.,* McGraw Hill, New York, NY. 1951.

Todd, David Keith: *Groundwater Hydrology*. John Wiley & Sons, New York, NY. 1959.

William, K. : *Irrigation Investment Technology, and Management Strategies for Development Studies in Water Policy Management 9*, Boulder, Colo Westview Press, 1986.

Index

A

Academy of Sciences, 37, 43, 44, 61.
Analysis of Water, 196.
Anthropologists, 94, 96, 97.
Anthropology, 92, 95.
Antimicrobial Resistance, 230.
Appropriative Rights, 102.

B

Biological Community, 33.

C

Challenges of Sustainable, 258.
Change Indicators, 33.
Chemical Analysis, 262.
Coastal Indicators, 25.
Coli Isolated, 230.

D

Data Analysis, 186.
Drainage Services, 149, 159.
Drinking Water, 200, 205, 206, 237, 258.

E

Effectiveness, 188, 189.
Electricity Prices, 88.
Electricity Water, 68, 70, 90.
Environmental, 14, 24, 25, 27, 30, 48, 67, 75, 90, 95, 174, 175, 177, 178, 179, 181, 182, 253, 262, 268, 269.
Environmental Condition, 27.
Established Water, 1.

F

Field Exercise, 182.
Field Sites, 176.

G

Global Dimensions, 117, 118.
Global Policy, 218, 220.
Greater Flexibility, 58.

H

Hydrology Courses, 177.

I

Implications, 20, 36, 52, 98.
Integrated Water, 218, 219, 220.
Integration of Hydrology, 174.
Irrigation Policy, 168.
Irrigation Water, 125.

L

Limitations of Current, 25.
Liturgical Praxis, 271.
Liturgical Theology, 269.

M

Management, 38, 45, 47, 48, 50, 58, 81, 100, 106, 110, 118, 149, 169, 210, 214, 218, 235, 236, 237, 272.

Management Scenarios, 45, 50.
Measurements, 179, 182.
Metropolitan Areas, 70.
Modelling Procedure, 130.
Modern Organisation, 196, 211.

N

New Generation, 22.

O

Outbreaks in Drinking, 242.

P

Photopigments, 31.
Policy Implications, 36, 52, 98.
Policy Narratives, 199.
Public Permits, 103.

Q

Quo Policy, 4, 14, 20.

R

Rare Resources, 222.
Recommendations, 50, 195, 201.
Responsiveness, 129.
Role of Agriculture, 119.

S

Sampling Strategy, 260.
Statistical Analysis, 233.
Susceptibility Testing, 232.

T

Thermoelectric, 79, 80.
Tougher Requirements, 56.
Transferability, 4, 182.
Treatment Plants, 231.

U

Undetermined Etiology, 247.
Uninterruptible Status, 43.

W

Wastewater, 192, 230, 231, 237.
Water Allocation, 41, 52, 58.
Water Crisis, 67, 79, 90, 91, 267, 271.
Water Disinfection, 236, 237.
Water Law, 3, 48.
Water Policy, 36, 52, 65, 90.
Water Quality Testing, 191.
Water Rights, 3, 36.
Water Scarcity, 117.
Watershed-based, 174, 182.

❑❑❑